Student Solutions Manual

Mathematics for Elementary School Teachers

FIFTH EDITION

Tom Bassarear
Keene State College

Prepared by

Ronald Yates
College of Southern Nevada

BROOKS/COLE
CENGAGE Learning

Australia • Brazil • Japan • Korea • Mexico • Singapore • Spain • United Kingdom • United States

ISBN-13: 978-1-111-56904-4
ISBN-10: 1-111-56904-5

Brooks/Cole
20 Davis Drive
Belmont, CA 94002-3098
USA

Cengage Learning is a leading provider of customized learning solutions with office locations around the globe, including Singapore, the United Kingdom, Australia, Mexico, Brazil, and Japan. Locate your local office at:
www.cengage.com/global

Cengage Learning products are represented in Canada by Nelson Education, Ltd.

To learn more about Brooks/Cole, visit
www.cengage.com/brookscole

Purchase any of our products at your local college store or at our preferred online store
www.cengagebrain.com

Printed in the United States of America
1 2 3 4 5 6 7 14 13 12 11 10

Student Solutions Manual: Table of Contents

CHAPTER 1 Foundations for Learning Mathematics

SECTION 1.1

1. 8 tricycles and 24 bicycles.

3. The three possibilities are:

Bicycles	Tricycles
6	0
3	2
0	4

5. 30 pigs and 139 chickens.

7. 100 five-dollar tickets and 500 two-dollar tickets.

9. You pay $16 for 5 movies, so you pay $3.20 per movie.

11. Sally makes $4.98 per hour..

13. **a.** If one penny has a diameter of 0.75 inches, then we have 216.525 billion inches, 18.04375 billion feet, 3,417,400 miles.

    b. 347 pennies per second. 10,950,000,000 pennies per year.

15. **a.** Need to make assumptions for the distance driven and the average cost of gasoline.

    b. Answers will vary. If the distance is 2400 miles and the cost of gas averages $1.50/gallon, then it will take $87.50 extra to drive the van.

    c. Answers will vary. Using the assumptions in part b, if the price of gas rises 40¢, then it is now averaging $1.90/gallon. Using the same procedure as in part b, it costs $53.33 more for the van and $30 more for the sedan.

17. **a.** 3600 beats per minute

    b. 216,000 times per hour

19. 36 toothpicks

1

SECTION 1.2

1. **a.** next term 21, 20th term 210

    b. next term 63, 20th term 104875

    c. next term 2, 20th term 10

    d. next term 18, 20th term 33

    e. next term 25, 20th term 400

3. **a.** $1 + (n-1)\, 3 = 3n-2$

    b. $7 + (n-1)\, 11 = 11n-4$

    c. $3*2^{(n-1)}$

    d. $2*3^{(n-1)}$

    e. $n\,(n+1)\,/\,2$

    f. $n\,(n-1) - (n-5)$ or n^2-2n+5

    g. n^2

    h. n^3

    i. $2^n - 1$

    j. $2^{(n+1)} - 2$ or $2\,(2^n - 1)$

5. Answers will vary.

7.

22	8	9	19
11	17	16	14
15	13	12	18
10	20	21	7

9.

a.	42	4662	The value of eah digit comes from adding two adjacent digits. Since it will be easier to understand later solutions by going from right to left, i.e., 1s place first, we will do that here. 2 4+2 = 6 4+2 = 6 4
c.	84	9324	The value of each digit comes from adding, as we did before. When the sums are greater than or equal to 10, we keep the digit from the 1s place of the sum and 'carry' the 1 from the 10s place to the next addition. The steps are now shown from right to left since this is how we determine the digits. 4 8+4=12 8+4+1=13 8+1=9
e.	441	48951	1 4+1=5 4+4+1=9 4+4=8 4

11.

14	17	20	23	26
10	13	16	19	22
6	9	12	15	18
2	5	8	11	14

13. There are lots of patterns. For example:
The sum of each row and each column is 260.
The sums of the diagonals are not 260; therefore, technically this is not a magic square.
If you decompose this figure into four 4×4 squares, the sum of each column and each row is 130.

15. a. The sum of the first seven terms is 33, which is one less than the ninth term. The sum of the first eight terms is 54, which is one less than the tenth term. So, the sum of the first n terms is equal to one less than the $(n+2)$nd term,
or $A_1 + A_2 + A_3 + \cdots + A_n = A_{n+2} - 1$

b. $A_n \times A_{n+2} = (A_{n+1})^2 - 1$.

17. Many possible answers.

19. a. There are many patterns. Next row is: 6 12 18 24 30 36

b. There are many patterns. Next row is: 1 13 61 129 129 61 13 1

c. There are many patterns. Next row is: 49 54 65 79 90 95 96

d. There are many patterns. Next row is: 3 21 63 105 105 63 21 3

21. 56, 62

3

SECTION 1.3

1. 26 rolls.

3. If you add two fractions that both have a 2 in the numerator, then the numerator of the sum is double the sum of the two denominators, and the denominator of the sum is the product of the two denominators.

   In notation: $\dfrac{2}{a} + \dfrac{2}{b} = \dfrac{2(a+b)}{ab}$.

5. **a.** If it is a fruit, then it contains sugar.

   b. If the mail carrier comes to the door, then the dog will bark.

   c. If the mail carrier comes to the door, then the cat will not bark.

   d. If it is a baby, then it cries at night.

   e. If you are to drive a car, then you must pass a written test.

7. **a.**

   b.

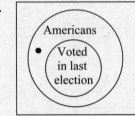

   c.

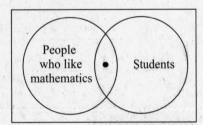

   d.

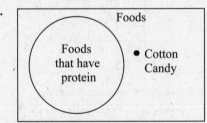

4

9. a. Invalid. Nothing is said about the relationship between professional athletes and rock stars. Although it is possible that some rock stars are professional athletes, it does not follow from the two statements.
One possibility:

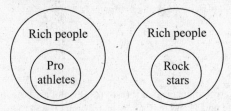

Another possibility:

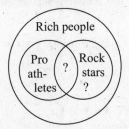

b. Invalid. There is more than one possible representation.

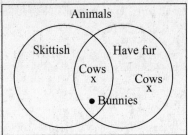

c. Valid

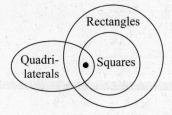

d. Valid

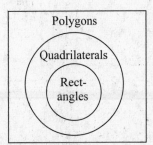

e. Invalid. The Venn diagram is not the best representation for this problem—and this is an important point.

11. a. The possible scores are 4 through 19, 21, 22, 24, 25, 26, 28, and 32.

b. Two ways: $1+1+2+8$ and $2+2+4+4$.

c. There must be an even number of ones to have the result be even. There are three ways to get an even number: no 1s, two 1s, or four 1s. Of course, the last possibility will not give us 12. If we use two 1s, then the remaining two throws must result in a total of 10. The only way to do this is $2+8$. Finally, using no 1s, suppose we have one 8 pt throw, then we must score the remaining 4 points in three throws, which is impossible. Thus an 8 pt dart cannot be used. If we have a 4-point throw, then the remaining three throws must result in 8 points. The only way to do this is $2+2+4$.

13. 64 games

15. We don't know, because an answer of 3:15 assumes that the power was off only a matter of seconds. It is possible that the power went off at 1 A.M. and came back on at 3:15 A.M. Thus we can say only that the power came back on at 3:15 because electric clocks begin at 12 A.M. when the power comes back on. If it is now 6:45 A.M., and the clock says 3:30, that means the power came back on $3\frac{1}{2}$ hours ago, that is, at 3:15 A.M.

17. 6 trips

19. B

5

SECTION 1.4

1. For all parts, let the bottom of the well be 0 feet.

    a. 9 hours.

    b. 23 hours.

    c. 19 hours

    d. 1 + [(a-b)/(b-c)]

        The brackets represent the greatest integer function, which means that if (a-b)/(c-d) is not a whole number, round that amount up the the next whole number and add 1, and that's how many hours it will take.

3. Bring over the goose (the fox will not eat the corn). Return and bring over the fox, but return with the goose. Bring over the corn. Return and bring over the goose again.

5. There are 160 triangles: 36 regular triangles, 28 inverted triangles, 28 regular , 15 inverted , 21 regular , 6 inverted, 15 regular , 1 inverted , and 10 regular triangles.

7. The ball bounced 4 times.

9.

44	5	5	5	2				
44	5	3	3	3	3			
44	5	3	3	2	2	2		
44	5	2	2	2	2	2	2	
44	3	3	3	3	3	2		
44	3	3	3	2	2	2	2	
44	3	2	2	2	2	2	2	2

11. 300 handshakes.

13. a. $\dfrac{104-(8\times 6)}{7}=8$-inch space between each plate and between the end plates and the wall.

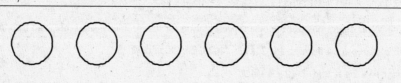

    b. $\dfrac{104-(8\times 5)}{7}=10\dfrac{2}{3}$ inches

    c. If she wants 12 inches between the end plates and the wall, then the space between each plate must be $\dfrac{104-(12\times 2)-(8\times 5)}{5}=6\dfrac{2}{3}$ inches .

15. a. There are three ways to get 42 and two ways to get 37.

    b. The scores that are impossible to get are: 1, 2, 4, 5, 8, and 11.

    c. 11 is the greatest score that is impossible to get.

CHAPTER 1 REVIEW EXERCISES

1. 316 student tickets

2. 16 stools

3. 15 coins equaling 92 cents

50	25	10	5	1
1	1	0	1	12
1	0	0	7	7
0	1	1	11	2
0	1	5	2	7
0	2	1	5	7
0	0	5	8	2

4. $216

5. 9 ways

6. 40 posts

7. $1.17

8. 43,200

9. There were two 10 cent stamps and ten 5 cent stamps.

10. The product is a four-digit number. The first two digits have a value one less than the number you are multiplying by 99. The second two digits equal what you get if you subtract that number from 100. Using symbols, if we let the number $= ab$, then the first two digits are $(ab-1)$ and the second two digits are $(100-ab)$.

11. P

12.

	Next term	20th term	nth term
a.	31	121	6n+1
b.	32	$2^{19} = 524288$	$2^{(n-1)}$
c.	242	$3^{20}-1$	3^n-1
d.	95	$(2^{20}*3)-1$	$(2^n * 3) - 1$

7

13. There are so many possibilities. Here are several:
    The magic sum is 34.
    The sum of the 4 numbers in the center is also 34.
    If you partition the 4×4 square into four 2×2 squares, the sum of the numbers in each of the 2×2
        squares is also 34.
    There are 2 odd and 2 even numbers in each row and column.
    The two middle rows and/or columns can be switched without affecting the square.
    The top-left to bottom-right diagonal has a constant difference of 3.
    If you look at the middle two columns, each pair of numbers contains consecutive numbers.

14. **a.** The sum of the numbers along the length of the stick equals the number at the end of the stick.

    b. Since the largest number on the chart is 924, work backwards: 1, 6, 21, 56, 126, 252, 462

    c. The handle would be the diagonal row of thirteen 1's along the left edge of the chart.

    d. Answers will vary.

15. **a.** Three ways

    b. 23 , 27, 29, 30, and 31 are impossible.

16. **a.**
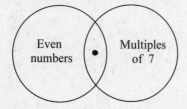
    b.
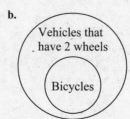

17. **a.** If it is meat, then it contains protein.

    b. If you want to pass the course, then you must attend class.

18. Inverse: If you don't work hard, then you won't succeed.
    Converse: If you succeeded, then you worked hard.
    Contrapositive: If you didn't succeed, then you didn't work hard.

19. This is invalid

20. This argument is invalid.

8

21. **a.** Fill the 9-gallon pail and use it to fill the 4-gallon pail. Empty that pail and fill it again from the 9-gallon pail. You now have 1 gallon left in the 9-gallon pail

      b. Fill the 4-gallon pail and empty it into the 9-gallon pail. Do it again. Fill the 4-gallon pail a third time and now fill the 9-gallon pail – it takes 1 more gallon to do so. You now have exactly 3 gallons left in the 4-gallon pail.

22. 11 hours

23. 26 packages can be made, with 2 ounces of seeds left over.

CHAPTER 2 Fundamental Concepts

SECTION 2.1: Sets

1. a. $0 \notin \varnothing$ or $0 \notin \{\ \}$

   b. $3 \notin B$

3. a. {e, l, m, n, t, a, r, y} and $\{x \mid x$ is a letter in the word "elementary"}
     or $\{x \mid x$ is one of these letters: e, l, m, n, t, a, r, y}.

   b. {Spain, Portugal, France, Ireland, United Kingdom (England/Scotland), Western Russia, Germany, Italy,
     Austria, Switzerland, Belgium, Netherlands, Estonia, Latvia, Denmark, Sweden, Norway, Finland, Poland,
     Bulgaria, Yugoslavia, The Czech Republic, Slovakia, Romania, Greece, Macedonia, Albania, Croatia,
     Hungary, Bosnia and Herzegovina, Ukraine, Belarus, Lithuania}.
     Also $\{x \mid x$ is a country in Europe}.

   c. {2, 3, 5, 7, 11, 13, 17, 19, 23, 29, 31, 37, 41, 43, 47, 53, 59, 61, 67, 71, 73, 79, 83, 89, 97}.
     Also $\{x \mid x$ is a prime less than 100}.

   d. The set of fractions between 0 and 1 is infinite.
     $\{x \mid x$ is a fraction between zero and one}.

   e. {name1, name2, name3, etc.}.
     $\{x \mid x$ is a student in this class}.

5. a. $\subset$ **b.** $\in$ **c.** $\subset$ **d.** $\subset$ **e.** True.

   f. False; red is an element, not a set.

   g. False; gray is not in set S.

   h. True.

7. a. 64

   b. A set with n elements has 2^n subsets.

9. a. Students who are members of at least two of the film, science, and computer clubs.
     $(F \cap S) \cup (S \cap C) \cup (C \cap F)$

   b. Students who are members of both the science and computer clubs, but not the film club.
     $\overline{F} \cap (S \cap C)$

   c. $\overline{C} \cap (S \cap F)$ **d.** $\overline{F \cup S \cup C}$

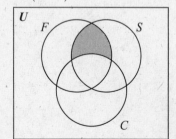

11. a. Students who have at least one cat and at least one dog.

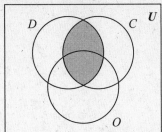

b. Students who have neither cats nor dogs.

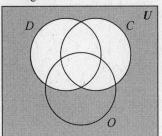

c. Students who have at least one cat, at least one dog, and at least one other pet.

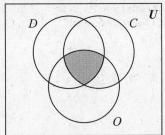

d. $D \cap \overline{C}$

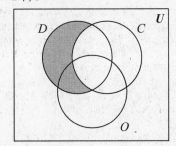

e. $D \cup C \cup O$

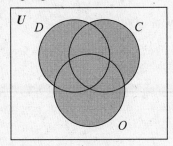

f. $O \cap (\overline{D \cup C})$

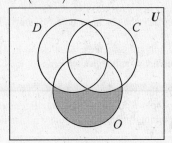

g. $\overline{D \cup C \cup O}$ or $\overline{D} \cap \overline{C} \cap \overline{O}$

Students who have no pets.

h. $C \cap (\overline{D \cup O})$

Students who have at least one cat and no other pets.

13. Answers will vary.

15. Answers will vary.

17.

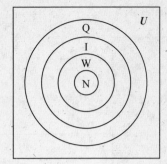

19. a.

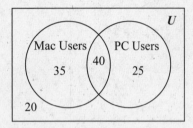

b. Yes, they are well defined.

21. a. A lesson in which the teacher would be using a lab approach with small groups.

b. Technically, this subset represents a lesson in which the students use concrete materials but not in a lab approach and not in small groups. Pedagogically this doesn't make sense.

23. a. Theoretically, there are four possibilities. I would pick the one at the left, because I think there can be successful people who are not very intelligent, intelligent people who are not successful, people who are successful and intelligent, and people who are neither.

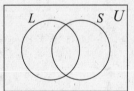 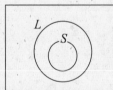 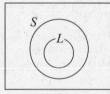 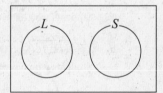

b. Answers will vary.

SECTION 2.2: Exercises

1. a.

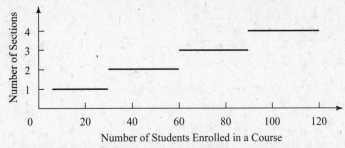

Student Enrollment and Course Sections

b. Yes. The number of course sections is determined by the number of students enrolled in the course. For a given number of students there is a specified number of sections.

3. a. 7 buses.

b.

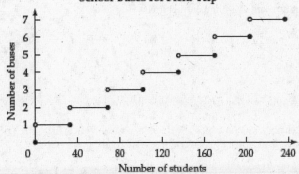

School Buses for Field Trip

5. No.

7. It would be less expensive to rent a car for a week.

9. Card 1 would be a better buy if you usually make a few long calls because the per-minute rate is cheaper. Card 2 would be a better buy if you usually make many short calls because there are no connection fees.

11. $2n^2 + 3n$

13. $12n - 3$

15. Either the first or third graph.

17. a.

Number of Elements	Number of Subsets
1	1
2	4
3	8
4	16
5	32

    b. If there are n elements in a set, that set will have 2^n possible subsets.

    c. $f(n) = 2^n$

19. Disagree. Functions can have two different inputs mapping to the same output, as long as there is only one possible output for each input.

21. a. 52.5 degrees

    b. No, because the y-intercept is not 0 (0 chirps corresponds to 40 degrees). In order for variables to be proportional, the equation must be of the form $y = ax$.

    c. $T = \dfrac{C}{4} + 40$, where T is temperature and C is number of chirps.

    d.

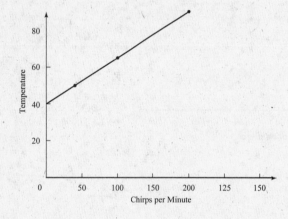

23. $3n + 2$

25. a. If we assume the surface area of one square on a block is 1 square unit, then a stack of n blocks has a surface area of $4n + 2$ square units.

    b. $6n + 4$

27. The boat can hold three possibilities: one adult, one child, or two children.

    a. 33 trips.

    b. $4x + 1$ trips

    c. 35 trips.

    d. $32 + 2y - 1$ trips.

    e. $4x + 2(y-1) + 1 = 4x + 2y - 1$

14

29. Answers will vary.

31. Answers will vary

33-35. Answers will vary.

37. *First graph:* The flag is raised at a steady rate.
Second graph: The flag is raised slowly at first, gradually increasing in speed.
Third graph: The flag is raised in small bursts.
Fourth graph: At first not much happens, then the flag is raised quickly almost to the top, then it takes a little time to reach the very top.

39. Answers will vary.

41. a. 8

    b. 16

43. C

45. E

SECTION 2.3: Exercises

1.

	Maya	Luli	South American
7		lokep moule tamlip	
8			teyente-toazumba
12			caya-ente-cayupa
13		is yaoum moile tamlip	caya-ente-toazumba
15		is yaoum is alapea	
16	uac-lahun	is yaoum moile lokep moile tamop	toazumba-ente-tey
21	hunkal-	is eln yaoum alapea	cajezea-ente-tey
22	hunkal-ca	is eln yaoum tamep	cajezea-ente-cayupa

3.

		Egyptian	Roman	Babylonian
a.	312	𓏲𓏲𓏲∩‖‖	CCCXII	𒀸𒀸𒀸𒀸𒀸 ⟨𒀸𒀸
b.	1206	⚡𓏲𓏲‖‖‖‖‖	MCCIIIIII or MCCVI	⟨⟨ 𒀸𒀸𒀸𒀸𒀸𒀸
c.	6000	⚡⚡⚡⚡⚡⚡	MMMMMM	𒀸 ⟨⟨⟨⟨
d.	10,000	𓂧	MMMMMMMMMM	𒀸𒀸 ⟨⟨⟨⟨𒀸𒀸𒀸𒀸𒀸 ⟨⟨⟨⟨
e.	123,456	𓆼𓂧𓂧⚡⚡⚡𓏲𓏲𓏲𓏲∩∩∩∩∩‖‖‖‖‖‖	Can't do	⟨⟨⟨𒀸𒀸 ⟨𒀸𒀸𒀸𒀸 ⟨⟨⟨𒀸𒀸𒀸 𒀸𒀸 𒀸𒀸𒀸 𒀸𒀸𒀸

5. a.   26

b.   540

c.   25

d.   450

e.   three thousand four hundred

f.   3,450

7. a.

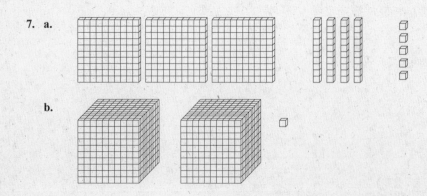

b.

9. a. 4859

 b. 30,240

 c. 750,003

11. a. 1004_{five} b. 334_{five} c. $0ff_{\text{sixteen}}$ d. 1101_{two}

 e. 1001_{two} f. $10f_{\text{sixteen}}$ g. 113_{four} h. 56_{seven}

13. a. $500+10+10+5+1+1=527$

 b. $100+50+10+10+5+1=176$

 c. ⲎⲎⲎΔΔΔΔΓ‖

 d. ⌈X⌉⌈H⌉⌈Δ⌉Γ

15. a.

 b.

 c. 460,859

 d. 135,246

17. a. 585 cartons of milk

 b. It has all 6 characteristics because this system is essentially base 6. The places are called cartons, boxes, crates, flats, and pallets. The value of each place is 6 times that of the previous place.

19. The child does not realize that every ten numbers you need a new prefix. At "twenty-ten" the ones place is filled up, but the child does not realize this. Alternatively, the child does not realize the cycle, so that after nine comes a new prefix.

21. Yes, 5 is the middle number between 0 and 10

23. We mark our years, in retrospect, with respect to the approximate birth year of Jesus Christ—this is why they are denoted 1996 A.D.; A.D. stands for Anno Domini, Latin for "in the year of our Lord."Because we are marking in retrospect from a fixed point, we call the first hundred years after that point the first century, the second hundred years the second century, and so on. The first hundred years are numbered zero (for the period less than a year after Jesus' birth) through ninety-nine. This continues until we find that the twentieth century is numbered 1900 A.D. through 1999 A.D.

25. Place value is the idea of assigning different *number values* to *numerals* depending on their position in a number. This means that the numeral 4 (four) would have a different value in the "ones" place than in the "hundreds" place, because 4 ones are very different from 4 hundreds. (That's why 4 isn't equal to 400.)

27. a. 11.57 days

 b. 11,570 days, or 31.7 years.

29. a. 21

 b. 35

 c. 55

 d. 279

     e. 26

     f. 259

     g. 51

     h. 300

     i. 13

     j. 17

     k. 153

     l. 1305

31. base 9

33. $x = 9$

35. This has to do with dimensions. The base 10 long is 2 times the length of the base 5 long. When we go to the next place, we now have a new dimension, so the value will be 2×2 as much. This links to measurement. If we compare two cubes, one of whose sides is double the length of the other, the ratio of lengths of sides is 2:1, the ratio of the surface area is 4:1, the ratio of the volumes is 8:1.

37. **a.**

     b.

     c.

39. Answers will vary.

41. B

43. C

CHAPTER 2 REVIEW EXERCISES

1. a. $\{x \mid x = 10^n,\ n = 1, 2, 3, ...\}$

 b. $\{10, 100, 1000, 10,000, ...\}$ or $\{10^1, 10^2, 10^3, ...\}$

2. No, because not all countries are recognized by all other countries.

3. a. $\in$ **b.** $\subset$ **c.** $\subset$ **d.** $\not\subset$

4. a. $D \not\subset E$ **b.** $0 \notin \{\ \}$

5. a.

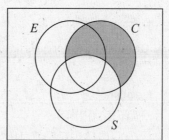

 b. $5, 15, 25$

 c. The set of even numbers between 0 and 30.

6. a.

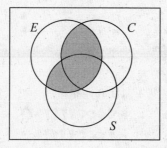

 b.

7. 50

8. The former means the same elements, and the latter means the same number of elements.

9. a.

x	$f(x)$	
0	-2	$0 \to -2$
1	1	$1 \to 1$
2	4	$2 \to 4$
3	7	$3 \to 7$

 b. 16

10. 28 rubles

19

11. $2(n+2)+1=2n+5$

12. $2(2n+1)+1=4n+3$

13. **a.** 110

  b. $n(n+1)$ gives the *n*th number.

14. **a.** 46.

  b. $4n+6$

15. If he averages 400 or more minutes per month, then use the second plan. Otherwise, use the first plan.

16. Graph (a) is not a match because the temperature of the coffee will not go below room temperature. Graph (b) is not a good match because it implies that the coffee cools at a steady rate. The coffee will cool more rapidly at the beginning. As it gets closer to room temperature, the cooling will get slower. Graph (c) is the best match because the rate of cooling slows down over time.

17.

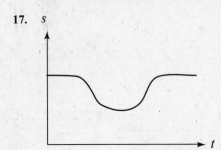

18. Answers will vary. **One story:** independent is time; dependent is distance. Jackie walked steadily for a period of time and then sat down.

19.

		Egyptian	Roman	Babylonian
a.	47	∩∩∩∩ |||||||	XLVII	⟨⟨⟨⟨ ▼▼▼▼▼▼▼
b.	95	∩∩∩∩∩∩∩∩∩ |||||	XCV	▼ ⟨⟨⟨ ▼▼▼▼▼
c.	203	99 |||	CCIII	▼▼▼ ⟨⟨ ▼▼▼
d.	3210	₤₤₤ 99∩	MMCCX	⟨⟨⟨⟨⟨ ▼▼▼ ⟨⟨⟨

20. **a.** 410_{five} **b.** 1300_{five} **c.** 1010_{two}

21. **a.** 4314_{five} **b.** 30034_{five} **c.** 1011_{two}

22. $25 \times 2 + 4 \times 5 + 3 = 73$

23. Because $1000_{five} = 125_{ten}$, I would rather have 200_{ten}.

24. They both have the value of 3 flats, 2 longs, and 1 single. Because base 6 flats and longs have greater value than base 5 flats and longs, the two numbers do not have the same value.

25. Because we are dealing with powers. Thus, the value of a base 10 flat is $2 \times 2 = 4$ times the value of a base 5 flat.

26. The value of the 5th place in base 10 is $10^4 = 10,000$. The value of the 5th place in base 5 is $5^4 = 625$.
  $10000 \div 625 = 16$.

27. There are many possible responses. Here are three: "One-zero" is the amount obtained when the first place is full. It means you have used up all the single digits in your base. It is the first two-digit number.

28. The visual representation of the seventh place is a cube.

29. There are several equivalent representations:

$2000 + 60 + 8$

$2 \times 1000 + 0 \times 100 + 6 \times 10 + 8 \times 1$

$2 \times 1000 + 6 \times 10 + 8 \times 1$

$2 \times 10^3 + 0 \times 10^2 + 6 \times 10^1 + 8 \times 10^0$

$2 \times 10^3 + 6 \times 10^1 + 8 \times 10^0$

30. Answers will need to include all six characteristics described in the section.

CHAPTER 3 The Four Fundamental Operations of Arithmetic

SECTION 3.1 Understanding Addition

1. **a.** $3+4=7$ **b.** $2+4=6$ **c.** $3\times2=6$

3. The equation for the set of ordered pairs (x, y) is $x+y=10$. In slope-intercept form it is $y=-x+10$, which is a straight line with a y-intercept of $(0, 10)$ and a slope of -1.

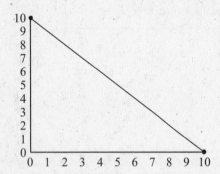

5. Exact answer: 787 miles

7. **a.**
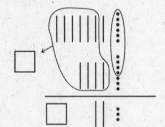
   b.

9. Answers will vary. When 6 and 7 are added, we are really adding $60+70=130$, so the 13 is shifted over because we have thirteen 10s. Then adding 3 and 5 we are really adding hundreds, so $300+500=800$ and the 8 goes in the hundreds column.

11. 5 hundreds + 10 tens + 13 ones = 5 hundreds + 11 tens + 3 ones = 6 hundreds + 1 ten + 3 ones = 613

13. **a.** After adding $6+8=14$, the student wrote the 1 in the ones place, carried the 4, and added it to $3+2$.

    b. Rather than "carrying," the student wrote the sum of $6+8$ in the ones and tens places and the sum of $3+2$ in the next available place.

    c. The student added the one from $5+7=12$ and the one from $6+8=14$ to the $3+2$ in the hundreds place; that is, the student seems to believe that the "carried" ones are carried until the end.

15.

$$\begin{array}{cccc} 5 & 5 & 6 & 8_{ten} \\ 2 & 7 & 4 & 5_{ten} \end{array}$$

0/7	1/2	1/0	1/3

$$\begin{array}{cccc} 8 & 3 & 1 & 3_{ten} \end{array}$$

$$\begin{array}{ccc} 3 & 2 & 2_{five} \\ 2 & 3 & 4_{five} \end{array}$$

1/0	1/0	1/1

$$\begin{array}{cccc} 1 & 1 & 1 & 1_{five} \end{array}$$

$$\begin{array}{ccc} 7 & 6 & 4_{eight} \\ 2 & 1 & 5_{eight} \end{array}$$

1/1	0/7	1/1

$$\begin{array}{cccc} 1 & 2 & 0 & 1_{eight} \end{array}$$

17. a. Base 8

b. Base 8.

c. The base for a is at least 4 and the base for b is at least 6.

Possible b's are: $7, 10, 13, \ldots$ and the corresponding a's would be $6, 8, 10, \ldots$

19. a.   37   56   74

b.   73   32   94

c.   24   47   97

d.   115   164   153

e.   185   153   274

f.   352   423   439

21. Not reasonable. Beginning with the 1000s place, we can see that we have 8000. Then going to the 100s place, we quickly see that there is more than 900.

23. $c = 9$ and $a + b = 7$, where a and b are integers

25. Some possibilities: $487 + 13, \ 479 + 21, \ 469 + 31, \ 468 + 32$

27. $N = 9, \ P = 2$.

SECTION 3.2 Understanding Subtraction

1. Answers will vary.

3. **a.** $7 - 5 = 2$ **b.** $7 - 5 = 2$ **c.** $8 \div 2 = 4$ (repeated subtraction)

5. **a.** $a - b \neq b - a$, because $b - a = -(a - b)$. For example, let $a = 7$ and $b = 4$. Then, $7 - 4 = 3$, and
    $4 - 7 = -3$.

    b. $(a - b) - c \neq a - (b - c)$, because $(a - b) - c = a - b - c$ and $a - (b - c) = a - b + c$. For example,
    $(10 - 4) - 3 \neq 10 - (4 - 3)$, because $6 - 3 \neq 10 - 1$.

7. *One possible way:* \$145,000 - \$116,000 = (\$145,000 - \$115,000) - \$1,000 = \$30,000 - \$1,000 = \$29,000

9. *One possible way:* $4132 - 2824$: $32 - 24 = 08$, $41 - 28 = 13$, $4132 - 2824 = 1308$ students

11. **a.** Samantha is using rectangles to represent 10s and dots to represent ones. When she writes out the part to be
     subtracted, she crosses those pieces off the first list, borrowing from the 10s symbols as needed.

     b. Alice is basically doing the same thing as Samantha only she writes one of the 10s rectangles as a sum giving
     $10 (4 + 6)$, so she can borrow the 6 and leave behind the 4.

13. **a.** $904 - 300 = 604$ The method works because
             $604 - 60 = 544$ $904 - 367 = 904 - (300 + 60 + 7)$
             $544 - 7 = 537$ $= 904 - 300 - 60 - 7$

     b. $367 + 7 = 374$ $904 - 367 = abc \implies 904 = 367 + abc$
             $374 + 30 = 404$ First he adds what is needed to get the c digit correct,
             $404 + 500 = 904$ carrying the extra to the 10s column. Then he
                                    follows by working on "balancing" the 10s and 100s
             $500 + 30 + 7 = 537$ column.

     c. $904 - 360 = 544$ This method is like the method in part (a), only the
             $544 - 7 = 537$ first two steps have been combined.

15. **a.**   Step 1: Regroup   Step 2: Takeaway 268

b.   Step 1: Regroup   Step 2: Takeaway 345

 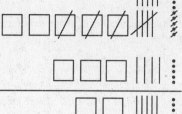

17. a.   The student subtracted, $8-6$. Some students automatically subtract the smaller number from the larger number, even if it changes the order of the numbers in the problem.

b.   The student wrote $6-8=2$, knew that 8 is bigger, and so "borrowed" from the tens place.

c.   The student renamed the 6 in the ones column as 16 before subtracting, but didn't change the 7 in the tens place to 6.

d.   The student probably reversed the numbers in the ones column in order to subtract, $8-0=8$, instead of renaming the 7 in order to subtract, $10-8$. Alternatively, the student could have reasoned that when you have a 0 in the minuend, you just bring down the digit in the subtrahend.

e.   The student changed the zero in the ones place to 10, but didn't rename the 7 as 6 in the tens place.

f.   The student renamed the 7 as 6 and each of the zeros as 10. The zero in the tens place should have been renamed as 9.

g.   The student does not realize they are borrowing the 10 from the larger place-value column. They know to do $14-9$, but do not change the 6 to 5 to compensate. Thus the next two column subtractions are wrong.

h.   The child is looking at the column $2-9$ and rearranging to actually do $9-2=7$. Since $3-2$ does not need rearranging in the child's eyes, they do $3-2=1$. The child does not understand that there is a difference between $a-b$ and $b-a$.

19.   Answers will vary. One possibility for each is given below.

a.   Leading digit: 2100

b.   Round both up to next thousand: $66,000-30,000$

c.   28 to 30 is 2, 30 to 72 is 42, so an estimate is 44,000.

d.   16 to 26 is 10, to 36 is thus 20, to 43 is thus 27, that is, 27,000.

e.   Think of $413-285$. 285 to 300 is 15, to 413 is 113 more. $113+15=128$. Thus, estimate is 128,000.

f. $18+14=32$. So 180,000 to 320,000 is 140,000. Can refine the estimate by taking off 4000 to say 136,000.

g.   Rounding: just over 1 million.

h.   Rounding: about 40 million.

21. a. There is more than one solution. b can be any digit, but $a = 0$, $c = 4$, and $d = 2$.

    b. $x = 2$, $y = 4$

23. Not reasonable. By using leading digit and looking at the hundreds place, we can quickly see that the answer will be less than 6000.

25. $3020_{five} - 441_{five} = 2024_{five}$

27. Some possibilities: $968 - 734$, $946 - 712$. There are many!

29. The mule was carrying 5 bales of cloth and the horse was carrying 7 bales.

31. The primary reason for the lack of tables is the inverse relationship between the two operations. If you know your addition and multiplication "facts," then you already know the corresponding subtraction and division "facts." For example, if you know that $4 \times 8 = 32$, then you know that $32 \div 8 = 4$, etc. For subtraction, I would say that a table is unnecessary - if you know your addition facts, you don't need a subtraction table. Thus, a subtraction table would send a mixed message that this is something new and important. A second reason is that a subtraction table would imply more complexity. Since subtraction is not commutative, there would be more "facts" to learn. For division, a table would not serve any purpose, there are no patterns.

33. 130 miles between Fresno and Bakersfield.

35. A

37. B [Add: 64% of 4th graders got this correct.]

SECTION 3.3 Understanding Multiplication

1. Answers will vary.

3. 4 x 3 because we are adding 3 four times.

5. No. Assign numerical values to *a*, *b*, and *c* to show that $a+(b \times c) \neq (a+b)(a+c)$.

7. **a.** The upper and lower halves are symmetrical, which illustrates the commutative property of multiplication.

   b. There are many ways to express the patterns. The most basic is that if you look at the two ends of the diagonal, you find the same number. As you move from the ends to the center of the diagonal, you continue to find matching numbers. Some diagonals have a center number with no partner and some do not. The diagonals with a center number have odd numbers at the ends.

   c. Both products of the diagonals have the same factors. If the factors of the top left term are *a* and *b*, the factors of the products of the diagonals are a, b, $a+1$, and $b+1$.

   Using algebraic notation, we can say:

	x	$x+1$
y	yx	$y(x+1)$
$y+1$	$(y+1)x$	$(y+1)(x+1)$

   d-e. See Exercise 9.

9. When nine is multiplied by a number *n*, the product can be thought of as ten times the number minus the number, $9n = 10n - n$. In this case, $9 \times 7 = (10 \times 7) - 7$. When *n* is 10 or less, the value of the digit in the tens place of the product is $10(n-1)$, and the value of the digit in the ones place is $9 - (n-1)$. Thus the sum of the digits is equal to nine. Since the bent finger represents *n*, the $n-1$ fingers to the left of *n* accurately represent the tens place, and the remaining fingers to the right of the bent finger represent the ones place.

11. 900 people

13. **a.**

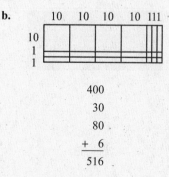

14 tens + 35 = 175

b.

$$400$$
$$30$$
$$80$$
$$+ \ 6$$
$$516$$

c.

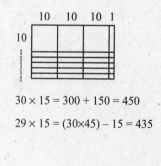

$30 \times 15 = 300 + 150 = 450$

$29 \times 15 = (30 \times 45) - 15 = 435$

15.

	30	4
20	600	80
8	240	32

$$\begin{array}{r} 600 \\ 80 \\ 240 \\ 32 \\ \hline 952 \end{array}$$

17. a.

$$\begin{array}{r} 72 \\ \times\ 34 \\ \hline 8 \\ 280 \\ 60 \\ 2100 \\ \hline 2448 \end{array}$$

$\leftarrow 4\times 2$
$\leftarrow 4\times 70$
$\leftarrow 30\times 2$
$\leftarrow 30\times 70$

This representation is nice because you don't have to do any carrying. Like using FOIL on $(70+2)(30+4)$.

b. This method is almost the same as the method in part (a); the order of multiplication is different.

c.

$$\begin{array}{r} 72 \\ \times\ 34 \\ \hline 8 \\ 28 \\ 6 \\ 21 \\ \hline 2448 \end{array}$$

$\leftarrow 4\times 2$
$\leftarrow 4\times 7$ (shifted because the 7 is in the tens place)
$\leftarrow 3\times 2$ (shifted because the 3 is in the tens place)
$\leftarrow 3\times 7$ (shifted because the 3 and 7 are both in the tens place)

19. **a.** Set up the numbers as shown at the right.

Find the four partial products and place them in the appropriate places: the ones digit in the lower spot and the tens digit in the higher spot of each cell. Note: If the product of the two numbers is less than 10, you can either leave the top cell blank or place a 0 in that spot.

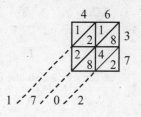

To find the product, find the sum of the numbers on each of the diagonal rows. If the sum of any diagonal row is above ten, "carry" the digit in the tens place to the next place.

b. One way to understand why the lattice method works is to compare it to obtaining the product using expanded form (shown at the right). Looking at both algorithms, you can see the placement of the partial products: 42, 28, 18, and 12. The lattice algorithm somehow enables each digit to be in its proper place.

$$40 + 6$$
$$\underline{30 + 7}$$
$$42$$
$$280$$
$$180$$
$$1200$$

You can verify the correct placement of each digit by examining the actual value represented by each partial product. That is, $6 \times 7 = 42$ and the 2 is in the ones place and the 4 is in the 10s place. Similarly, 6×3 represents 6×30, which is 180. In the traditional algorithm, we don't even write the 0 (we just move over). Similarly, in the lattice algorithm, we deal only with "significant" digits: The 8 must be in the tens place and the 1 must be in the hundreds place.

c. You must know the multiplication table (how to multiply one-digit numbers) and how to add a list of one-digit numbers and "carry" if necessary. You must also be familiar with the ones and tens digit places in base 10.

d. Answers will vary.

21. Rather than carrying the regroupings above the multiplication (where they are added in immediately following the multiplication step), the student places them in the addition rows (where they are added in during the addition step). She does this so that the multiplication and addition are completely separated.

23. The child put the 8 from 28 below and then also carried the 8.

25. Answers will vary.

27. b is wrong. Estimate it as $20 \times 900 = 18,000$, which is not even close to the answer, $30,102$.

29. 3720 miles

31. The value of the largest place contributes more to the final sum than all of the other places combined.

33. **a.** A novice would say reasonable because 60 x 80 = 4800 and thus is under 4800, but more sophisticated students would say that, thinking of the four partial products, it will be clearly less than 4798.

b. No. This answer essentially takes the smallest and largest of the partial products: 8×4 and 4×3. The answer is also off by one full place: $400 \times 300 = 120,000$—and that error is the larger of the two.

35. a. 72 x 49. Reasoning 82 x 49 would be close to 80 x 50 = 4000 which is too big.

b. There are several possible ways to get 5 in the ones place. If we look at partial products and begin with a 'middle possibility' shown below we can see that we have 5400 + 450 + 300 + 25 which is going to be about 6200.

$$90 + 5$$

$$x\ 60 + 5$$

So we can try a bit smaller. This gives us 5400 + 450 + 180 + 15 = 6045.

$$90 + 3$$

$$x\ 60 + 5$$

c. A quick estimate enables to know that the value of the 10s place of the second number is 4. Thus we have 50 something times 43 = 2193. We can get the same answer by dividing 2193 by 43.

d. A quick glance tells us that we are looking at 80 something x 50 something and the answer has a 1 in the ones place and we need 611 from the partial products. What numbers can give us a 1 in the ones place? We try 83 x 57 or 87 x 53.

37. A ream of paper is typically 500 sheets of paper.

a. 600,000 sheets of paper

b. $18,000

c. $12,000 savings.

39. Answers will vary.

41. a. About 1 per hour, so that means about 24 per day; round to 25 since it is slightly more than 1 per hour. $25 \times 365 = 9125$ per year.

b. Answers will vary. Some possibilities: How did they find out? Do they test both driver's and passenger's blood for alcohol when there is a fatal accident? Does this imply the drinker is the driver of the car or in the car? Does this count all kinds of traffic accidents: motorcycle, car hits pedestrian?

43. a. 3,796,416

45. a. This shortcut will work for any two digit numbers that begin with the same number where the sum of the second two digits is 10.

b. Label the digits of the two numbers to be multiplied: $ab \times ac$. The last two digits of the result come from multiplying $b \times c$. The digits that precede this come from $a \times (a+1)$.

c. $ab \times ac = (10a+b)(10a+c)$
$$= 100a^2 + 10ab + 10ac + bc$$
$$= 10a(10a+b+c)+bc$$
$$= 10a(10a+10)+bc$$
$$= 100a(a+1)+bc$$

The 100 in the product $100a(a+1)$ simply moves the term two places to the left, thus, the product $a(a+1)$ gives the first part of the answer and bc gives the second part.

47. **a.** The ones place of the missing multiplier is an 8. Without knowing the product, that is all we can say.

    b. 68×83 will be above 5000 and 78×83 will be above 6000, so the multiplier is 68.

    c. 8×83 and 18×83 will both be less than 2000, so the missing multiplier is either 8 or 18.

49. **a.** $530 \times 71 = 37,630$

    b. $357 \times 01 = 357$.

51. **a.** 425   **b.** 1872   **c.** 5040

    d. I don't have a complete justification. I know that we are adding the same numbers as would be added in the Egyptian duplation method. However, I don't know why you add only the numbers in the "Doubles" column whose partner in the "Halves" column is an odd number.

53. **a.** $(30+4)(20+3) = 600+90+80+12$

    $(3x+4)(2x+3) = 6x^2+9x+8x+12$

    If x represents the number 10, then $6x^2+9x+8x+12 = 600+90+80+12$.

    b. The FOIL algorithm works because it specifies the four partial products.

55. Alike: sum (product) of any row, any column, both diagonals are identical.

    Difference: multiplication magic square, don't seem to need all numbers in the square to be different.

57. This is a game to be played by students so there are no answers.

59. 7

SECTION 3.4 Understanding Division

1. 42 jelly beans. The partitioning model.

|||| : |||| : |||| : |||| :

3. a. $3 \times 2 = 6$ (repeated addition)

    b. $8 \div 2 = 4$ (repeated subtraction)

5. a. 300 b. 80 c. 50 d. 300 e. 40 f. 50

7. Answers will vary.

9. There are multiple ways to get answers.

    a. 6 x 10 = 60 and 6 x 5 = 30 which gives an estimate of 15.

    b. 450/15 = 30 and 45/15 = 3, so a quick estimate is 30; a refined estimate is 33

    c. 180/6 = 30 and 18/6 = 3, so a quick estimate is 30; a refined estimate is 33

    d. Cancel the zeros and we have 26/2 which is 13. Here the exact answer is easily obtained.

    e. 500/25 = 20 and 75/25 = 3, so a quick estimate is 20; a refined estimate is 23

    f. 750/25 = 30 and 100/25 = 4, so a quick estimate is 30 and the exact answer is 34

    g. Cancel the zeros and we have 240/5. Since 240/10 = 24, 240/5 = 48.

11. 4 cans.

13. $700 per month.

15. a. 3023_{five} b. 203_{five} r4_{five} or $(203 \frac{1}{2})_{five}$

17. Exact answer: 675 cases per week and 2700 cases per month

19. Round up to get 53 weeks.

21. 89 days

23. a. The quotient must be greater than 25.

25. Not reasonable because 60 x 60 = 3600 and if you think about the partial products (shown below), you can see that their value is clearly more than 400.

$$60 + 7$$

$$x \quad 60 + 8$$

27. Melanie needs to sell 160 tapes.

29. Suppose each glass of juice contains 8 ounces; then each guest will get 16 ounces of juice. Wei needs to buy 60 containers of juice. The juice will cost $53.40.

31. One. As I was going to St. Ives . . . everyone else was going the other way!

33. Answers will vary. Assumptions that must be made include the size of the page, the size of the type, the number of columns per page.

35. 3 million crimes per year would mean about 17,000 crimes per school day (dividing by 180 school days per year). If we divide 17,000 by 50 states, that's about 340 crimes per day per state. That still seems high, so the term "crime" must be broad: vandalism, minor assault. Does it mean "crimes reported to police"?

37. Not quite. At this rate, she will burn exactly 370 calories.

39. A carton contains $4 \text{ packs} \times \dfrac{6 \text{ sodas}}{\text{pack}} = 24 \text{ sodas}$. 247 students divided by 24 sodas per carton is 10 remainder 7. Round up to get 11 cartons total.

41. 1961, 30, 48, 20,000, 20

43. a. They will make the trip in less than 30 days.

    b. Exact answer: 5010 miles

45. Exact answer: $238.19

47. (c) is wrong, the actual quotient is 78.

49. Francie is correct. There are many ways to arrive at this conclusion. Thinking of partitioning, if you increase the whole (dividend) and decrease the number of groups (divisor), each group will get more. If you increase the whole, then you've got to increase the divisor too.

51. Each child gets 4_{five} gumdrops.

53.

    a.

a	b	$a+b$	$a \cdot b$
65	66	131	4290
1208	72	1280	86,976

    b.

a	b	$a+b$	$a \div b$
153	9	162	17
19	1	20	19

    c.

a	b	$a-b$	$a \cdot b$
44	24	20	1056

    d.

a	b	$a \cdot b$	$a \div b$
10	2	20	5
34,225	37	1,266,325	925

55. a. $2\overline{)974} = 487$

    b. $9\overline{)247} = 27.4$

57. 42857

33

59. a. One solution is $66 + 80 - 3$.

    b. One solution is $23 \times 60 + 23 \times 10$.

    c. One solution is $(260 \div 10) \div 2$.

    d. One solution is $25 \times 44 + 1 \times 44$.

    e. One solution is $653 + \mathbf{150} = 803$ and $803 + \mathbf{31} = 834$.

61. a. $[(A + B + C) \times D] / 100$

    b. $(A + B) \times C$

    c. $A + BC - D - E$

63. D

65. A

67. B

69. He accidentally multiplied.

CHAPTER 3 REVIEW EXERCISES

1. **a.** $7+2$ **b.** $7-4$ **c.** 2×3 **d.** $12\div3$

2.

    0 1 2 3 4 5 6 7 8 9 10 11 12 13 14 15 16 17 18 19 20

3. Responses will vary. This algorithm shows the sum of each place, beginning with the largest place. Once the sum for each place is written down, it is relatively easy to add those numbers mentally to write down the sum.

4. $345-97$.

5. We have simply changed the representation of the minuend from 8 hundreds to 7 hundreds + 9 tens + 10 ones.

6. Responses will vary.

7. Responses will vary. At the heart of the matter is that, regardless of the places involved, 10 of something is being exchanged for 1 of something else.

8. Responses will vary. A common solution path is to see the four partial sums and add them: 6 hundreds, 14 tens, 12 tens, and 28 ones.

9. The digits being multiplied to get the second row are 37×2. However, this represents 37×20, which is 740. We commonly omit the zero and just write 74; however, because its real value is 740, we must place the 7 and the 4 in the correct places.

10. Responses will vary. At the heart of the matter is repeated addition. Essentially, we have 13 groups of 25. We can more easily get the answer by finding 10 groups of 25 (250) and 3 groups of 25 (75) and adding these amounts together.

11. 541×82. Responses will vary.

12. Responses will vary. One response is to note that a visual representation of the problem involves four partial products, only two of which are shown here. Pete's method gives us only two of the four products needed.

13. Responses will vary. The heart of the response is to note that if we are adding, when we take 1 from 17 and give it to 29, it is literally 1. However, if we are multiplying, when we take 1 from 17 and give it to 29, we are really taking one group of 29, rather than just 1.

14. Responses will vary. At a concrete level, using the example cited, the origins of 12×20 are (3×4) and (4×5). The origins of 15 and 16 are (3×5) and (4×4). Using the commutative and associative properties, we can show the equivalence of (3×4) and (4×5) and (3×5) and (4×4).

35

15. $a = 16$ and $b = 30$ or vice versa.

16. Using a calculator, $73932500 \div 97665 \approx 757.00097$. Multiplying 757×97665 gives us 73932405, which we can subtract from 73932500 to get 95. Thus $x = 757$ and $y = 95$.

17. The answer is an infinite sequence beginning with 91 and increasing by 75 each time. That is, 91, 166, 241, 316, and so on.

18. 12

19. $79 \times 9 \times 11 = 7821$.

20. In each case, we are determining how much is being used up by the digit in the divisor. That is, when we say 5 goes into 43 eight times, we are saying that we are using up $80 \times 5 = 400$.

21. Responses will vary.

22. Responses will vary.

23. a. Commutative and associative properties transform the problem into $(36 + 64) + 82$, which equals $100 + 82$.

b. The distributive property transforms the problem into 20×19.

c. Representing 1592 as $1600 - 8$, and then using the distributive property of division over subtraction transforms the problem into $1600/8 - 8/8$.

24-28. Responses will vary. One possible response for each is presented here.

24. a. $3684 + 8312 \approx 3700 + 8300 = 12,000$
$2853 + 6241 \approx 2900 + 6200 = 9100$
$12000 + 9100 = 21,100$ for an estimate

b. $44268 - 28843 \approx 44000 - 29000 = 45000 - 30000 = 15,000$ for an estimate.

c. $468 \times 9 \approx 468 \times 10 = 4680$, which is an overestimate by 468. So $4680 - 470 = 4210$ is an estimate.

d. $48 \div 14 \approx 3\frac{1}{2}$, so multiplying by 100, I would estimate 3500.

25. $\$345,300 - \$216,250 \approx \$345,000 - \$215,000 = \$130,000$

26. She makes approximately 30 trips per semester and drives about 125 miles each trip. $30 \times 125 = 375$ miles.

27. $489 \div 19 \approx 500 \div 20 = 25$ miles / gallon

28. Round each price to the nearest 50¢: $\$1.50 + \$2.50 + \$2.00 + \$0.50 + \$3.50 + \$1.00 + \$1.00 + \$1.00 = \$13.00$

29. When we change 638×42 to 638×40, we are decreasing the answer not by 2 but by 638×2, roughly 1200. Thus, when we round up 638, we need to make up that 1200. If we round 638 to 640, we are increasing by 2×42; if we round 638 to 660, we are increasing by 22×42.

30. 338 days

31. 3 times

32. **a.** 1 ream = 500 sheets ; 1200 x 500 = 600,000 pages

   b. $18,000

   c. $12,000 (2 cents per page)

33. **a.** 11330_{five} **b.** 2211_{five} **c.** 3113_{five} **d.** 3214_{five}

34. Base 7.

35. Base 8

36. Base 4.

37. **a.**
$$\begin{array}{r} 3\;{}^{1}6\;{}^{1}8\;4 \\ +4\;2\;4\;8 \\ \hline 7\;9\;3\;2 \end{array}$$

   b.
$$\begin{array}{r} {}^{5}\cancel{6}\;\cancel{0}\;\cancel{0}\;3 \\ -d\;2\;8\;4 \\ \hline 2\;7\;1\;9 \end{array}$$

CHAPTER 4 Number Theory

SECTION 4.1 Divisibility and Related Concepts

1. Answers will vary.

3. A number that is not evenly divisible by 2.

    A number that ends in 1, 3, 5, 7, or 9

5. **a.** Divisible by 3, 4, 6 and 8

    b. None of these numbers divide 2,345,678

7. **a.** True. a only needs to be a factor of b or c for $a \mid bc$.

    b. False. 24 is divisible by 3, but neither 2 nor 4 is divisible by 3. If every digit of a number is divisible by 3, then the number is divisible by 3.

    c. False. $4 \mid 2 \cdot 6$, but $4 \nmid 2$ and $4 \nmid 6$.

    d. False. $4 \mid 36$ and $6 \mid 36$, but $24 \nmid 36$.

    e True. 6 is a factor of 12. If $a = 12n$, then $a = 6 \times 2n$.

    f. True. If $a = dn$, then $a^2 = d^2 n^2$.

9. **a.** 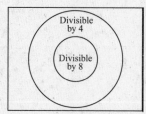 **b.** 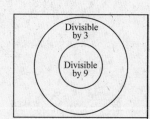 **c.**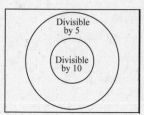

    d. 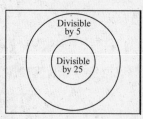 **e.**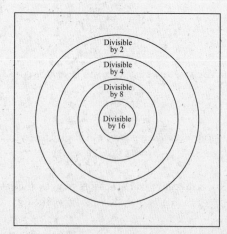

11. The sum of these numbers will be odd.

   If there is an odd number of odd numbers, then the answer will be odd. If there is an even number of odd numbers, then the answer will be even. The number of even numbers does not affect the sum in terms of being even or odd.

13. a. Recalling the diagram of an odd number from the text, we can conceptualize each odd number as an L shape; that is, two equal lengths and then one dot that joins the two sides. Visually, we see that each successive square number can be decomposed to a smaller square and an L shape. The big idea here is not formal proof but the notion of composition and decomposition.

   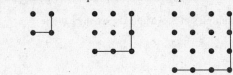

   b. Once again, composition and decomposition. Turn the triangle numbers on their side:

   Each triangle number and the following triangle number can be assembled into squares. The figure below at the left shows $1 + 2 + 3 + 2 + 1$ as $(1+2+3)+(2+1)$. The figure at the right shows $1+2+3+4+3+2+1$ as $(1+2+3+4)+(3+2+1)$.

   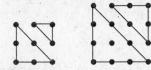

15. Zero can be divided in two equal pieces too, zero and zero.

17. a. No.

   b. No.

   c. Yes.

   d. Yes.

19. a. The sum of three consecutive numbers will always be divisible by 3 and by the middle number in the sequence.
   $$x + (x+1) + (x+2) = 3x + 3 = 3(x+1)$$

   b. The sum of four consecutive numbers will always be divisible by 2.
   $$x + (x+1) + (x+2) + (x+3) = 4x + 6 = 2(2x+3)$$

   c. The sum of five consecutive numbers will always be divisible by 5 and the middle number in the sequence.
   $$x + (x+1) + (x+2) + (x+3) + (x+4) = 5x + 10 = 5(x+2)$$

21. This could be misinterpreted. For example, from this description, someone could conclude that 8 does not divide 324 because 2 is not divisible by 4.

23. **a.** The number must be divisible by 2 and 9.

  b. The number must be divisible by 3 and 8.

  c. The last two digits must be 00, 25, 50, or 75. If we list multiples of 25, we quickly see the pattern.

  d. There are several possibilities. One is that the sum of the digits must be an even number. The most elegant is that the number is divisible by 2 if there is an even number of even digits in the number.

  e. The sum of the digits must be divisible by 4. This is an adaptation of the divisibility rule for 9 in base 10.

  f. The last digit must be 3, 6, 9, or 0. If we list multiples of 3, we quickly see the pattern.

25. The answer is 79.

27. 61

  If you begin to look at possible answers, you see patterns:
  If 4 players, possibilities are 5, 9, 13, 17, 21, 25, 29, 33, 37 …
  If 5 players, possibilities are 6, 11, 16, 21, 26, 31, 36, 41 …
  If 6 players, possibilities are 7, 13, 19, 25, 31, 37, 43, 49 …
  We quickly see that for both 4 and 6, the ones place is 5, 9, 3, 7, or 1 and for 5 players, the ones place is always 6 or 1. Thus, after our list we only need to try 51, 61, 71, etc.

29. **a.** 30 and 15

  b. 12 and 36

  c. There are many possibilities: 12 and 49, 13 and 48, 14 and 47, 15 and 46, 16 and 45, 17 and 44, 18 and 43, or 19 and 42.

31. **a.** No. Since 4 does not divide 46, 4 does not divide 4,346.

  b. Yes. Since the sum of the digits is 21, 3 divides this number, and since 4 divides 68, 4 divides this number therefore, it is divisible by 12.

  c. No. Since the sum of the digits is not divisible by 3, 3 does not divide this number and therefore 15 does not divide this number.

33. One solution—Joan's ages: 61, 62, 63, 64, 65, 66

  granddaughter's ages: 1, 2, 3, 4, 5, 6

35. **a.** The last digit is 1, beacuse the exponent is divisible by 4.

  b. The remainder is 2, because the last digit is 7.

  c. The remainder is 1.

37. **a.** 1, 3, 7, 9 or 1, 3, 5, 11

  b. 9

  Combine $1+3$ with pairs of odd numbers whose sum is 26. Use $1+5$ and pairs of odd numbers whose sum is 24. Then use $1+7$ and $3+5$ with $9+13$. Finally, combine $3+7$ with $9+11$. Going from smallest to largest odd numbers in a systematic way ensures that all possible combinations are listed.

39. **d.**

SECTION 4.2 Prime and Composite Numbers

1. Any prime number has exactly 2 factors.

3. **a.** $2\times2\times2\times2\times3$

   b. $3\times5\times5$

   c. $2\times2\times23$

   d. $2\times2\times2\times2\times3\times3$

   e. $2\times2\times7\times7$

   f. $2\times2\times2\times3\times3\times7$

   g. $2\times2\times3\times3\times3\times7$

   h. $2\times3\times11\times13\times17$

5. Prime: 521, 523

   Immediately cross out even numbers and numbers ending in 5 or 0.

   511: 7×73

   513: sum of digits is divisible by 3

   517: 11×47

   519: sum of digits is divisible by 3

7. a. 1, 2, 3, 4, 6, 8, 9, 12, 18, 24, 36, 72 Solution paths will vary.

   b. 1, 2, 3, 5, 6, 10, 15, 25, 30, 50, 75, 150

9. a. We know that odd x odd is odd. Since 3^2 and 5^2 are both odd, their squares will be odd. When we multiply their product (odd) by 2^3 (even), we will get an even number.

   b. Even. Same reasoning.

11. **a.**   1, 3, 7 (factors of 21)   **b.**   1, 2, 3, 4, 6, 9, 12, 18 (factors of 36)

13. Answers will vary. Some possibilities are 16, 81, and 625.

15. Dolls with numbers that are perfect squares (1, 4, 9, 16, 25, . . .) will be facing up at the end.

17. 121. It must be the square of a prime number.

19. 5, 25, and 125 are possibilities.

21. All numbers greater than 5 with a 5 in the ones place are divisible by 5, so the remaining candidates for consecutive prime triplets must have 7, 9, and 1 in the ones places or 9, 1, and 3 in the ones places. However, among these possibilities, one of the three numbers will always be divisible by 3.

23. Answers will vary.

25. B

SECTION 4.3 Greatest Common Factor and Least Common Multiple

1. **a.** 15   **b.** 9   **c.** 1   **d.** 27

   e. 3   **f.** 3   **g.** 26   **h.** 42

3. **a.** 60   **b.** 150   **c.** 132   **d.** 6930

5. Answers will vary.

7. 90 and 126, or 18 and 630

9. 24 and 300

11. 180 days (this is the LCM of 18, 12, and 15)

13. **a.** True.   **b.** True.   **c.** False, because it is not always true. Ex: $GCF(12,18) = 6$.

    d. True.   **e.** True.

15. Answers will vary. One possiblility is $120\,\text{cm} \times 120\,\text{cm} \times 120\,\text{cm}$.

17. **a.** 2×2 square   **b.** 6×6 square   **c.** The GCF of x and y.

19. **a.** The purple rod – that is, a rod 4 units long.

    b. The yellow rod – that is, a rod 5 units long.

    c. A train whose length is the GCF of x and y.

21. C

CHAPTER 4 REVIEW EXERCISES

1. **a.** Yes, the sum of the digits is divisible by 3.

   b. Yes; the number is divisible by both 2 and 3.

   c. No, the number represented by the last two digits is not divisible by 4.

2. Here are three equivalent statements:
   4 is a factor of 12.
   12 is a multiple of 4.
   12 is divisible by 4.

3. 4320

4. False. Counterexample: 6 divides 24 and 8 divides 24, but $6 \times 8 = 48$ does not divide 24.

5. **a.** 11 is one answer.   **b.** 107 is one answer.

6. 91, 93, 94, and 95 are the possible answers.

7. If the sum is odd, then one is even and one is odd. Therefore, their product must be even.

8. It will have a 1 in the units place.

9.
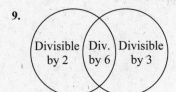

10. **a.** $2 \times 2 \times 2 \times 3 \times 3$ **b.** $2 \times 2 \times 3 \times 3 \times 7$

11. It is a prime number.

12. 49

13. Both 3 and 5 must divide *m*.

14. **a.** 8   **b.** 8   **c.** 4

15. **a.** 90   **b.** 120

16. 54 and 60

17. 54.

18. The two numbers must be relatively prime.

19. 192 or 216.

CHAPTER 5 Extending the Number System

SECTION 5.1 Integers

1. **a.** -218 **b.** -2 **c.** 19 **d.** -78 **e.** 14
   f. -18 **g.** 7 **h.** 2 **i.** 221 **j.** -6

3. $-5°$ Farenheit.

5. $864

7. $-$19

9. $-40°$ Celsius

11. ***One way:*** $-19 + -6 = -25$; $-25 + -22 = -47$; $-47 + 8 = -39$; $-39 + -4 = -43$; $-43 + 7 = -36$; and $-36 + 1 = -35$.

    Another way: $(-19 + -6 + -22 + -4) + (8 + 7 + 1) = -51 + 16 = -35$.

13. $2\dfrac{2}{3}$ hours (2 hours 40 minutes) difference in flight time.

15. Answers will vary. Here are some possibilities.

a.

18	7	11
5	2	3
13	5	8

b.

-5	-3	-2
-1	-1	-1
-3	-2	-1

c.

3	5	-2
8	8	0
-5	-3	-2

d. The difference in the far right column is $(a-b)-(c-d)$.

   The difference in the far left column is $(a-c)-(b-d)$.

a	b	$a-b$
c	d	$c-d$
$a-c$	$b-d$	$a-b - c+d$

17. Answers will vary.

44

19. **a.** Always positive. The absolute value of a nonzero expression is always positive.

     b. It depends. If $x > y$, then it will be positive. If $x < y$, then it will be negative.

     c. Always positive. $x^2 + y^2 > xy$

     d. It depends. If $x^2 + 2xy > y^2$, then it will be positive.

21. 176 pounds.

23. B

SECTION 5.2 Fractions and Rational Numbers

1. Answers will vary. Some possibilities include:

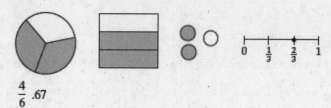

$\dfrac{4}{6}$ .67

3. larger piece = 1/3, smaller piece = 1/15

5. **a.** ¾   **b.** ¼   **c.** 2/5

   d. 1/10   **e.** ¼   **f.** 1/16

7. **a.** 2/10, 3/15, 4/20

   b. 6/8, 9/12, 12/16

   c. 4/6, 6/9, 8/12

   d. 10/12, 15,18, 20/24

9. **a.** 5/8   **b.** 2/3   **c.** 9/10   **d.** 2/5   **e.** 7/9   **f.** 7/11   **g.** 29/30   **h.** 1/5

11. **a.** The value of the regions from greatest to least is 1/2, 1/3, 1/6.

   b. The value of the regions from greatest to least is 9/24, 1/4, 1/6, 1/8, 1/12.

   c. Answers will vary.

13.

15. **a.** 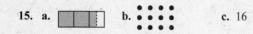   **b.**   **c.** 16

17. **a.** 3/7 < ½ < 5/8

   b. 5/6 < 9/10 Each 1 piece away from 1

   c. 2/7 < 1/3 < 4/11

   d. 7/9 < 15/17 Each 2 pieces away from 1

   e. 2/9 < ¼ < 5/16

   f. ¾ = 75/100 < 79/100

19. If you round both numbers down slighty, you have 35,000 / 50,000 ≈ 7/10.

21. **a.** 5/8

   b. 5/6

   c. 37/158 is between 1/5 and 1/4 . The middle thermometer is in that range.

23. 1/6. Even though there are five sections, they are not equal in size. The shaded portion is 1/3 of 1/2 of the box.

25. This is a valid, though unconventional, response. If you take a whole and divide it into 2 equal parts and then shade in $1\frac{1}{2}$ of them, you have the same area as if you had divided the whole into 4 parts and shaded in 3 of them.

   Alternatively, the ratio $1\frac{1}{2} : 2$ is equivalent to $3 : 4$.

27. **a.**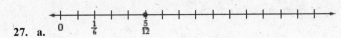

   b.

29. **a.** About ¾ of ¼ a tank, so ¾ * ¼ = 3/16

   b. About ½ of a tank plus another ½ of ¼ of a tank, so ½ + ½ * ¼ = ½ + 1/8 = 5/8

31. $\frac{10}{13}$ is closer to 1 than to $\frac{1}{2}$. $\frac{1}{2}$ of 13 is $6\frac{1}{2}$, so $\frac{1}{2} = \frac{6\frac{1}{2}}{13}$. $\frac{10}{13}$ is $\frac{3\frac{1}{2}}{13}$ more than $\frac{1}{2}$ and $\frac{3}{13}$ less than 1.

33. **a.** 5/15, 3/4, 4/5. 5/15 is less than 1/2; 3/4 and 4/5 are greater than 1/2. 3/4 is 1/4 less than 1; 4/5 is 1/5 less than 1. Since 1/4 is greater than 1/5, 3/4 is less than 4/5.

   b. 5/11, 2/3, 6/7. 5/11 is less than 1/2; 2/3 and 6/7 are greater than 1/2. 2/3 is 1/3 less than 1; 6/7 is 1/7 less than 1. Since 1/3 is greater than 1/7, 2/3 is less than 6/7.

   c. 3/50, 1/3, 2/5, 5/8, 3/4. 3/50 is very small compared to the others. 1/3 = 2/6. Since 1/6 is less than 1/5, 2/6 is less than 2/5, so 1/3 is less than 2/5. 5/8 is 1/8 more than 1/2; 3/4 is 1/4 more than 1/2. Since 1/4 is greater than 1/8, 3/4 is greater than 5/8.

   d. 3/10, 2/5, 4/7, 5/6, 7/8. 2/5 = 4/10, so 2/5 is greater than 3/10. 4/7 is slightly more than 1/2; 5/6 and 7/8 are close to 1. Since 5/6 is 1/6 less than 1 and 7/8 is 1/8 less than 1, and 1/6 is greater than 1/8, 5/6 is less than 7/8.

35. **a.** 2/3 and 3/4 are reasonable answers.

   b. 3/10   **c.** 3/4   **d.** 3/5

37. Approximately 1/5 of the trip is left. $24/115 ≈ 22/115$, which simplifies to 5/23. $5/23 ≈ 5/25$, which simplifies to 1/5.

39. Fuller. There are several ways to justify this. One way: How much larger (multiplicatively) is the denominator? That is, 32 times what = 264 and 58 times what = 402? Mentally, we can determine that $32 \cdot 8 = 256$ and $58 \cdot 7 = 406$.

   That is, $\frac{32}{264}$ is slightly less than 1/8 and $\frac{58}{402}$ is slightly greater than 1/7. So $\frac{58}{402}$ is larger.

41. a. There are 48 students (30 girls and 18 boys) in the chorus.

 b. 1/4 of the students.

43. a. He was 12 and I was 48. In 12 years, he will be 20 and I will be 60; in 28 years, he will be 40 and I will be 80.

 b. There are many solutions: ... (10, 40), (11, 44), (12, 48), (13, 52), (14, 56)...

 c. There are many answers. One is that both numbers are multiples of 10.

45. If the original fraction is less than one, the new fraction is larger. The difference between numerator and denominator becomes less significant as they increase, so they will be proportionately closer together, making for a larger fraction.

47. a. 5/16 **b.** 19/37 **c.** 199/307

 d. As x gets larger and larger, the value of the function gets closer to 2/3

49. D

51. B

SECTION 5.3 Understanding Operations with Fractions

1. **a.** $80\dfrac{17}{24}$ **b.** $-2\dfrac{23}{120}$ **c.** $23\dfrac{13}{20}$ **d.** $191\dfrac{11}{16}$ **e.** $-3\dfrac{7}{12}$

   f. $66\dfrac{7}{24}$ **g.** 99   **h.** 135   **i.** $13\dfrac{1}{2}$ **j.** 4 ½   **k.** 3/8

3. **a.** They added the numerators, but instead of first getting a common denominator, they just multiplied the denominators.

   b. They added the denominators.

   c. They added the denominators and changed to a mixed number using a base 10 idea, 13 = 1 whole and 3 parts, instead of 1 whole (8/8) and 5 parts.

   d. They subtracted the larger fraction from the smaller fraction, ignoring the order.

   e. They simply subtracted the numerators and denominators.

   f. $7\dfrac{1}{8}$ should equal $6\dfrac{9}{8}$ not $6\dfrac{11}{8}$.

   g. They multiplied 2×4 and 3×3, then added.

   h. They multiplied both the numerator and the denominator by 5.

   i. They cross-multiplied, one numerator by the other denominator.

   j. They divided the numerators and divided the denominators.

   k. They took the reciprocal of the first fraction instead of the second.

   l. They divided each part separately; $8\div2=4$ and $\dfrac{1}{8}\div\dfrac{1}{4}=\dfrac{1}{2}$.

5. **a.** 23/6   **b.** 23/4

7. Repeated subtraction. Since 5/5 = 1, we repeatedly subtract 5/5 from 13/5 until our remainder is less than 5/5.

9. Diagrams will vary.

a. $\frac{2}{3} \times 2\frac{3}{4}$:

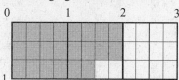

Rearranging the shaded area:

Total area $= 1 + \frac{2}{3} + \frac{2}{12} = 1\frac{5}{6}$

b. $2\frac{2}{3} \times 3\frac{1}{2}$:

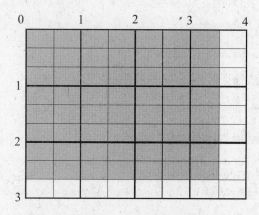

$= 6 + $ Fractional pieces

We have 6 whole pieces. Rearranging the fractional pieces:

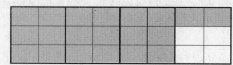

$= 3 + \frac{2}{6} = 3\frac{1}{3}$

Total area $= 6 + 3\frac{1}{3} = 9\frac{1}{3}$.

11. a. Less than 10. $\frac{5}{8} + \frac{3}{8} = 1$ and $\frac{3}{42} < \frac{3}{8}$.

b. Less than 2. One possible way: $\frac{3}{4} + \frac{1}{16} < 1$, so the sum of the three is less than 2.

c. Greater than 2. Since $\frac{1}{4} > \frac{1}{10}$, $\frac{1}{4} + 1\frac{9}{10} > 2$.

d. Between $5\frac{1}{2}$ and 6. $2\frac{2}{3} < 3$. Thus, $8\frac{1}{2} - 2\frac{2}{3} > 8\frac{1}{2} - 3 = 5\frac{1}{2}$.

e. Greater than 20. $8\frac{1}{2} \times 3$ (round up, round down) $= 25\frac{1}{2}$. Alternatively, $8 \times 2 = 16$, $8 \times \frac{7}{8} = 7$; we are already past 20.

f. Greater than 1/2. If we double $4\frac{7}{8}$, we will clearly be over 9.

13. 114 teachers.

15. 5/42 of the weight is additives. Let 1 = total weight.

Subtract the weight of the water and juice: $1 - \left(\frac{5}{7} + \frac{1}{6} \right) = 1 - \left(\frac{30}{42} + \frac{7}{42} \right) = 1 - \frac{37}{42} = \frac{5}{42}$.

17. a. $213\frac{1}{3}$ ounces left.

b. $17\frac{7}{9}$ glasses.

19. a. The child is adding the fractions as if they are pieces of a pie. The child took $\frac{1}{2}$ from $2\frac{1}{2}$ and added it to $\frac{1}{4}$. This gave 2 wholes plus $1\frac{1}{4}$. The child took apart $2\frac{1}{2}$ into $2 + \frac{1}{2}$ and then used the associative property to connect $\frac{3}{4}$ and $\frac{1}{4}$.

b. The child is taking fractions apart and putting them back together and using the associative and commutative properties: $2\frac{1}{2} + \frac{3}{4} = \frac{1}{2} + 2 + \frac{3}{4} = \frac{1}{2} + 2\frac{3}{4} = \frac{1}{4} + \frac{1}{4} + 2\frac{3}{4} = \frac{1}{4} + 3 = 3\frac{1}{4}$

21. a. 3/4 is 3 pieces of a 4-piece pie. Take 1/2 of each of the three pieces. The half pieces are each 1/8 of the pie, so three of them would be 3/8.

b. 3/4 is 3 pieces of a 4-piece pie. Divide two of the three pieces so that you keep one and give one away. Divide the last of the original three pieces in half, keep one and give one away. What you have left is $1/4 + 1/8 = 3/8$.

23. It is not just repeated addition. The most general model of multiplication is that it represents the area of the square formed by the two numbers. If you make a 1×1 square and find 1/4 of that square and then find 1/2 of 1/4 of that square, then you have 1/8 of the square.

25. a. *One way:* $26 \times 11 \div 12$. *Another way:* $11 \div 12 \times 26$.

b. $26 \times 12 \div 11$.

27. a. If $4\frac{1}{2} \times 60 = 270$, then $4\frac{1}{2} \times 15 = 67\frac{1}{2}$ and $4\frac{1}{2} \times \frac{1}{2} = \frac{2}{14}$. So the answer is $67\frac{1}{2} + 2\frac{1}{4} = 69\frac{3}{4}$.

b. If we double both numbers, we have 9×31. The answer to $4\frac{1}{2} \times 15\frac{1}{2}$ will be $\frac{1}{4}$ of 9×31. If $9 \times 30 = 270$ and $9 \times 1 = 9$, then $9 \times 31 = 279$. One-half of $279 = 139\frac{1}{2}$ and one-half of $139\frac{1}{2} = 69\frac{3}{4}$.

c. $15 \times 36 = 540$, so $15 \times 18 = 270$ and $15 \times 9 = 135$ and $15 \times 4\frac{1}{2} = 67\frac{1}{2}$. Now we need one-half of $4\frac{1}{2}$, which is $2\frac{1}{4}$. So $15 \times 4\frac{1}{2} + \frac{1}{2} \times 4\frac{1}{2} = 69\frac{3}{4}$.

29. The conceptual error is that the student does not realize $\dfrac{a+b}{c+d} \neq \dfrac{a}{c}+\dfrac{b}{d}$.

To demonstrate: $\dfrac{9\frac{1}{4}}{3\frac{3}{4}} = \dfrac{9+\frac{1}{4}}{3+\frac{3}{4}} \neq \dfrac{9}{3}+\dfrac{\frac{1}{4}}{\frac{3}{4}}$, rather, $\dfrac{9\frac{1}{4}}{3\frac{3}{4}} = \dfrac{9}{3\frac{3}{4}}+\dfrac{\frac{1}{4}}{3\frac{3}{4}}$.

31. **a.** $6 \div \dfrac{3}{4} = 6 \times \dfrac{4}{3} = \dfrac{24}{3} = 8$ guests can be served.

b. Thinking of each pint in 4 parts, there are 6×24 parts to split up. Giving 3 to each person, $24 \div 3 = 8$ guests.

c. $\dfrac{3}{4} \times ? = 6 \Rightarrow \dfrac{3}{4} \times ? = \dfrac{24}{4} \Rightarrow \dfrac{3}{4} \times \dfrac{8}{1} = \dfrac{24}{4}$, thus $\dfrac{3}{4} \times 8 = 6$.

d. Giving 3/4 to each guest, we'll add $\dfrac{3}{4}+\dfrac{3}{4}+\dfrac{3}{4}+...$ until we get 8.

e. We have 6 pints and we'll give away 3/4 pint repeatedly until we run out.

33. **a.** D **b.** A **c.** B **d.** E

35. 20 packages can be made, with 1 ounce of seeds left over.

37. 13 boxes.

39. The assumption is of monogamy. That is, the number of married women = the number of married men. Thus, there are more men than women in this community. One solution path: since 2/3 of the women = 1/2 of the men, draw a diagram to represent this fact. (Shown at the right.) Now, all the parts are the same size. Since there are seven parts in the whole, 3/7 of the population is single.

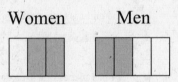

41. It will take 4 pressings to get at least 3/4 of the juice, and 6 pressings to get 9/10 of the juice.

43. **a.** $48

b. Answers will vary.

45. $\dfrac{8}{9}-\dfrac{6}{7}$

47. $\dfrac{7}{8} \div \dfrac{1}{9}$ if you assumed proper fractions; $\dfrac{8}{2} \div \dfrac{1}{9}$ if you did not.

49. **a.** 110_{five}

b. 20_{five}

51. D

53. 4 gallons

55. b

SECTION 5.4   Beyond Integers and Fractions: Decimals, Exponents, and Real Numbers

1. a.

    b.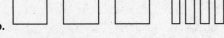

    c.

    d.

    e.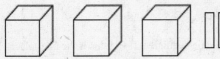

    f.

    g. 1.64

    h. 0.164

    i. 2.05

    j. 0.205

3. a. $4 + \dfrac{6}{10} = 4\dfrac{6}{10}$ b. $\dfrac{7}{10} + \dfrac{5}{100} = \dfrac{75}{100}$

    c. $1 + \dfrac{2}{10} + \dfrac{3}{100} + \dfrac{4}{1000} = 1\dfrac{234}{1000}$ d. $4 + \dfrac{6}{100} = 4\dfrac{6}{100}$

5. a. $0.\overline{6}$; 2/3   b. $0.\overline{5}$; 5/9   c. $0.\overline{09}$; 1/11   d. $0.\overline{142875}$; ½

7. a. 0.4   b. 0.98899   c. 0.05   d. 0.087

9   a. 0.0084, 0.058, 0.56, 0.6

    b. 0.0086, 0.065, 0.9, 1.04

11. $24,440,000 or $24.44 million

13. a. 0.067

b. 4060.034

15. a. 8.24   **b.** 16.804   **c.** 1.85   **d.** 1.244

     e. 11.18   **f.** 16.5435   **g.** $26.\overline{6}$ **h.** 24

17. a. 4.1   **b.** 0.64   **c.** 0.027   **d.** - 0.135   **e.** 0.03

     f. 42   **g.** 5.6   **h.** 16   **i.** 31.2   **j.** 0.5

     k. 2   **l.** 22.1   **m.** 0.609   **n.** 0.0452

19. a. Answers will vary. One possibility: 0.9995.

b. Answers will vary. One possibility: 3.445.

21. Since the 10ths place is the digit $\div$ 10, the oneths place would be the digit $\div$ 1. But this is the ones place. Thus the ones place and oneths place are the same place.

23. There are many answers.

     a. $3276 = 36 \times 91$, so $32.76 = 3.6 \times 9.1$.

     b. $476 = 2 \times 238$, so $4.76 = 0.2 \times 23.8$.

     c. $36 = 6 \times 6$, so $0.36 = 0.6 \times 0.6$.

25. a. We want 0.23 of $0.8\,\text{m}^3$. The word "of" is a good indication of multiplication: $0.23 \times 0.8 = 0.184\,\text{m}^3$.

     b. $\dfrac{40\ \text{miles}}{\text{gallon}} \times 0.75\ \text{gallons} = 30\ \text{miles}$. Using multiplication, the units cancel properly to give us miles.

     c. $5\ \text{liters} \times \dfrac{0.2\ \text{liters}}{\text{cup}}$ cannot be correct because the units do not cancel. Therefore this is a division problem:

         $5\ \text{liters} \times \dfrac{\text{cup}}{0.2\ \text{liters}} = 5 \div 0.2 = 25\ \text{cups}$.

     d. $7.2\ \text{pounds} \times \dfrac{\text{box}}{0.8\ \text{pounds}} = 7.2 \div 0.8 = 9\ \text{boxes}$

     e. $75\ \text{roses} \div 5\ \text{roses per bouquet} = 15\ \text{bouquets}$

     f. $\dfrac{1\ \text{yard}}{\$15.00} \times \dfrac{1}{0.65\ \text{yards}}$ will not work because the $ units are left in the denominator instead of the numerator.

         Thus, we try $\dfrac{\$15.00}{\text{yard}} \times 0.65\ \text{yard} = \$9.75$.

g. We want the answer to be in pounds per person: $5 \text{ pounds} \div 12 \text{ people} = \dfrac{5}{12}$ pound of cookies per person.

h. $\dfrac{16 \text{ miles}}{\text{sec}} \times 0.85 \text{ sec} = 13.6$ miles

i. $13.9 \text{ meters stretched} \times \dfrac{1 \text{ meter original}}{3.3 \text{ meters stretched}} = 13.9 \div 3.3 \approx 4.2$ meters original length.

27. Answers will vary.

29. 102 is the number of vials that can be completely filled. 0.4 is the fraction of another vial that can be filled. 102 vials will use 127.5 ounces. The remaining 0.5 ounce is 0.4 of a 1.25 ounce vial.

31. $632

33. $70.64

35. **a.** 1.23456789×10^8

    b. 3×10^{15}

    c. 5.6×10^{-10}

    d. 3.02×10^{-7}

37. $12,090,000,000,000, which is approximately $40,000 for every person in the United States (based on a population of 300 million).

39. Not much. Assume the average citizen drives 12,500 miles a year and has a car that gets 25 miles per gallon. The person would buy 500 gallons. $0.05/gallon × 500 gallons = $25.

41. **a.** The length of each candle.

    b. Answers will vary.

43. **a.** Light travels at a speed of 186,000 miles per second. To find out how far it travels in a year, you would do the following calculation:

        (186,000 miles/second)(60 seconds/minute)(60 minutes/hour)(24 hours/day)(365.25 days/year)

    b. (14 × 3356 × 789) × 10,000,000,000 = 37,070,376 × 10,000,000,000 = 370,703,760,000,000,000.

45. I guess they figure readers will be confused if they report it accurately, for example, $8\dfrac{1}{3}$ as 8.3 and $8\dfrac{2}{3}$ as 8.7.

47. 21/60 = 0.35, so it is 7.35 minutes.

49. **a.** 3.75

    b. 6.67

    c. 7:12 P.M.

    d. 8 A.M.

51. **a.** Diagram will be equivalent to 1/5.

    b. A0.A.

53. a. 4.8

    b. 4.0

55. E

57. D

CHAPTER 5 REVIEW EXERCISES

1. **a.** −4   **b.** 1/4

2. **a.** −74   **b.** 19   **c.** −6   **d.** −18

3. −2

4. It means $(-3)+(-3)+(-3)+(-3)+(-3)+(-3)$.

5. The first symbol represents an operation. The second symbol signifies the value of the number. To mix them up is to become sloppy, which is not a good habit to get into. The number sentence given reads, "Negative 3 minus 4 is equal to negative 7."

6. Yes. The whole has been divided into four regions of equivalent value, and three of those four regions are shaded. The horizontal line is extraneous.

7. Because $\dfrac{5}{6}=\dfrac{35}{42}$ and $\dfrac{6}{7}=\dfrac{36}{42}$, you can go to 84ths and see that $\dfrac{71}{84}$ is between the two.

8. There are three spaces between 0 and $\dfrac{1}{5}$, thus each space is $\dfrac{1}{5}\div 3=\dfrac{1}{5}\times\dfrac{1}{3}=\dfrac{1}{15}$. Counting by 15ths from 0 to the x, we can see that x is $11/15$.

9.   or

10.

11. Answers will vary. Figure should have four dots.

12. **a.** Both are two parts from 1. Because 11ths are greater than 15ths, 2/11 will be farther from 1, so 13/15 is greater.

    b. 7/12 is greater. Using 1/2 as a benchmark, we can quickly see that 7/12 > 1/2 and 13/28 < 1/2.

13. When we divide 1 by 2 in base 5, we get $0.2222..._5$.

14. It means they have the same value.

15. There are many ways to express the reason. One is to say that, by definition, the numerator of a fraction means how many "equal" pieces one has. If the denominators are not identical, then the numerators represent pieces of different size.

16. There are many ways to express why the hypothesis is invalid. One is to liken it to multiplication of two-digit numbers, such as 73×21. If you just multiply 70×20 and 3×1, you are missing the "two longs" regions.

17. We can solve it without the algorithm by finding $7\frac{3}{4}$, $2\frac{1}{3}$ times – that is,

$$7\frac{3}{4}+7\frac{3}{4}+\frac{1}{3}\left(6\frac{3}{4}+1\right)=7\frac{3}{4}+7\frac{3}{4}+2\frac{1}{4}+\frac{1}{3}=18\frac{1}{12}$$

18. It will be in region B because the answer is clearly positive, but it will be less than either of the two fractions because of the operator model of fractions.

19. She can make 15 cakes and will have $1\frac{1}{4}$ cups of flour left over.

20. You could do it in either of two ways: $4\frac{3}{4}$ is $\frac{1}{2}$ of $9\frac{1}{2}$, so this problem is more than $\frac{1}{2}$. Or $\frac{1}{2}$ of $8\frac{7}{8}$ would be $4\frac{3.5}{8}$, and because $\frac{3}{4}$ is greater than $\frac{3.5}{8}$, the answer is greater than $\frac{1}{2}$.

21. Answers will vary.

22. **a.** 1200 is a reasonable rough estimate, because 2/3 of 18 is 12, and 1754 is close to 1800. Using the distributive property, we can get closer: $\frac{2}{3}(1800-46)=1200-30=1170$.

    b. About 63 1/3 square inches.

23. **a.** 0.625   **b.** 0.004   **c.** 6/25

24. The number of valid answers is infinite. They include 0.44441, 0.44442, and so on.

25. -0.54 0.045 0.454 $5/11$

26. 23.3. The two nearest tenths are 23.2 and 23.3.

27. $4,306,000,000

28. **a.** $3.4\times10^{10}\times5.6\times10^{7}=3.4\times5.6\times10^{10}\times10^{7}=19.04\times10^{17}=1.904\times10^{18}$

    b. $5.82\times10^{5}\div2.3\times10^{10}=\dfrac{5.82\times10^{5}}{2.3\times10^{10}}=\dfrac{5.82}{2.3}\times10^{-5}=2.53\times10^{-5}$

29. **a.** 0.112 cubic meters. We multiply because we have part of the tank: fraction as operator.

    b. 53 cups. We divide because the problem is repeated subtraction – repeatedly taking away 0.15 liter.

    c. 8 boxes. Same reasoning as in part (b).

    d. $6.06. This can be seen as division as a proportion: $4.85/0.8=x/1$.

CHAPTER 6 Proportional Reasoning

SECTION 6.1 Ratio and Proportion

Notes: (1) All of these problems can be solved in more than one way; some of the alternative solution paths are noted. (2) This is not an algebra course; most of these problems can be solved with proportions, but there are no problems for which solving a proportion is the only solution path.

1. 143.75 calories.

3. 165 dentists.

5. a. $750   **b.** $1250   **c.** $977.08

7. 860,000 people.

9. Approximately 51 feet, 5 inches.

11. Yes.

13. $11\frac{3}{4}$ inches. Approximate answer.

15. a. 48 cents   **b.** 96 cents   **c.** $4.41

17. Answers will vary.

19. a. 0.15 lb   **b.** 14 oz or 0.875 lb   **c.** 49 million

21. Amy's red blood cell count is high.

23. 10 questions

25. 3.5 feet.

27. 37 handshakes per minute, which is one handshake in less than 2 seconds.

29. 2520 students.

31. $1\frac{1}{2}$ hours more.

33. More prison inmates.

35. The ratio between boys and girls becomes greater, because the ratio of boys to girls added to the class, 1:1, is greater than the original ratio, 3:8.

37. a.   The proportions of the ingredients vary with the quantity, due to the cooking process.

   b.   8 cups liquid, 5 cups cereal, 1/2 teaspoon salt. Answers may vary.

39. 199/325 = 61.2% and 5/8 = 62.5% These percentages are close enough to say that the advertisement is accurate. Also, 199/325 ≈ 200/325, which simplifies to 8/13, but 5/8 sounds better than 8/13 for advertising purposes.

41. The person who gets paid twice a month will have a larger paycheck.

43. If you assume that the truck slowed her down by 20 miles per hour and Sonja claims that she would have arrived at class 20 minutes earlier, then she followed the truck for $6\frac{2}{3}$ miles. It seems unlikely that she would not have been able to pass the truck in this distance, unless she was driving on a narrow, winding road.

45. a. Answers will vary. For example, there are more than 100,000 more Black males in prison than white males.

    b. Answers will vary. For example, the rate of incarceration for black males is more than 7 times the rate of incarceration of white males.

    c. Answers will vary. For example, from the raw numbers, the numbers for white females and black females are close. However, the rates for black females is almost 5 times the rate for white females. The disparity or inequity becomes more visible with rates.

47. a. Answers will vary. For example, the numbers in South Africa are slightly more than the number for India.

    b. Answers will vary. For example, while the numbers for South Africa and India are similar, the rate in South Africa is more than 20 times the rate in India.

    c. Answers will vary.

49. 52%, 62%

51. D

SECTION 6.2 Percents

1.	**a.**	18	**b.**	19.55	**c.**	1211.8	**d.**	54.4	**e.**	0.8625
	f.	240	**g.**	80%	**h.**	85.7%	**i.**	8.52%	**j.**	26.1%
	k.	18.75%	**l.**	145.8%	**m.**	46.35	**n.**	480	**o.**	7.192
	p.	70	**q.**	32.4	**r.**	43.65	**s.**	1185.6	**t.**	160

3. 35%. Using compatible numbers, $23 \times 3 = 69$, so 23 is just over 33%.

5. 700 patients. One way to estimate is to use guess–test–revise. 30% of 1000 is 300, so try smaller. Using this method, you can get the actual answer.

7. **a.** 131%. Estimating: 42,000 to 84,000 is 100% increase. Answer is more than 100%.

   b. 26%. Estimating: Using compatible numbers, 15,000/60,000 = 1/4.

   c. 43%. Estimating: Increase is about 31,000, or just under 1/2.

   d. 132%. Estimating: Similar strategy as in (a).

9. Approximately 6%. The baby lost 1/2 pound. $\dfrac{1}{2} \div 8 = \dfrac{1}{16} \approx \dfrac{6}{100}$. Adult would have lost approximately 10 pounds. If you see $6\% \approx 1/16$, then $1/16 \times 160 = 10$.

11. About 12.8 million. 12,812,800

13. Approximately 41% of eligible voters aged 18–24 voted in 2008.

15. $2200

17. $963.64

19. $30,941.04 after 2 years.

21. $22,373,000

23. About 1.2 miles.

25. The "whole" each year is not the same.
Let's take someone making $40,000. If they get a 6% raise per year, their salary after four years would be $50,499.08. If they got 12% in the first year and then 4% for the next three years, their salary after 4 years would be $50,393.91. However, and here is where math is not simple, the total four year's salary under the first plan would be $185,483.70 compared to $190,241.60 under the second plan. So after four years, they would have more total salary under the second plan but their base salary would be slightly higher under the first plan.
   At a simpler level, think of $40,000 with a 50% raise the first year and then no raise the second year vs. a 25% raise each year. In the first plan, your salary jumps to $60,000 and then stays at $60,000 the second year. Under the second plan, your salary jumps to $50,000 the first year and then it goes up 25% of $50,000 the second year and so it goes up to $62,500.

27. **a.** The high school principals make 93% more than the teachers; the junior high school principals make 81% more than the teachers; and the elementary school principals make 70% more than the teachers.

   b. The junior high school principals make 6% less than the high school principals. The elementary school principals make 12% less than the high school principals. The teachers make 48% less than the high school principals.

29. a. 5 out of 1000, or 1 out of 200, children will have a severe reaction.

b. Answers will vary. Possible questions include the following: What are the risks of not having the child vaccinated? How does the reaction compare with the illness itself? Are some children more likely than others to have a serious reaction?

31. 7% grade means that a hill changes 7 vertical feet for every 100 horizontal feet, or 70 vertical feet for every 1000 horizontal feet, or a proportionate amount of change. The percent grade gives drivers an indication of the steepness of a hill.

33. a. $4180.80 per year.

b. The raise was less than the rate of inflation. $1200/$23,400 $\times$ 100 $\approx$ 5% raise.

35. If the cost of the item is less than $100, take $10 off. If the cost of the item is greater than $100, take 10% off.

37. a. $10.29

b. $20,580

c. This assumes that she works 40 hours a week and 50 weeks a year.

d. She would be making $10.33 an hour, which is about $80 more a year.

39. $121,885.98

41. 12 years.

43. $2\frac{1}{2}$ hours

45. a. Using leading digit to get the total number of cars, we get about 140 million and 13/140 is approximately 1/10 or 10%

b. 9.8%

c. These data are probably available from the Registry of Motor Vehicles, which every state has. I would guess these data are pretty reliable.

47. Impossible to say without knowing how many people total attended each game.

49. a. $48 \text{in} \div 3\frac{3}{4} = 12$ with 3 in remaining

b. $\frac{3}{48} \times 100 = 6.25\%$ waste

c. 6.25% of $40,000 = $2,500 wasted

d. A 45-inch coil would result in no waste (although there are certainly other correct answers)

e. Both coils result in less waste than the 48-inch coil.

f. Answers will vary. It is essential to explain that since $48 \div 3.75 = 12.8$, the 0.8 represents 8/10 of the 3.75 in that is needed.

51. E

CHAPTER 6 REVIEW EXERCISES

1. $69.29

2. $1227.50

3. The ratio will increase.

4. They would need to add 7 teachers.

5. 559 miles

6. 50 more males

7. The birth rate is a ratio. Thus, to keep the ratio the same, if the population increases, the number of births also needs to increase. If the rate is down, it means that the number of births did not increase at the same rate as the population did.

8. We use rates to enable us to compare "apples and oranges." For example, let's say two cities had the same number of cases of a disease, but the second city had five times as many people as the first. If we get just raw numbers, we don't see the difference. However, if we convert the numbers to rates, the rate for the first city is five times the rate for the second, which tells us that the disease is much more prevalent in the first city.

9. Still need to raise 15% of the goal, which is $240,000.

10. To keep the same taste, she has to add both juices according to the ratio, which, in simplest terms, is 2 cups of grape juice for every 3 cups of orange juice. Thus, if she added 10 cups of grape juice, she would need to add 15 cups of orange juice to keep the ratios the same.

11. $22,373,000

12. 67% weight gain

13. The task will be complete in one more hour.

14. $\dfrac{86,400 \text{ sec}}{342.25 \text{ miles}} \approx 252 \text{ sec/mile} = 4.2 \text{ min/mile} \text{ or } 4 \text{ min } 12 \text{ sec/mile.}$

15. $1,575,000

16. 66% of the students have pets.

17. 4.6% of the world's population lived in the U.S. in 2002.

18. 30 students

19. 80%

20. 2.3%.

21. I had punched 0.0055 instead of 0.055.

22. Technically, the answer is $125.64, but it is more likely that the original price was $126. When you take 30% off $126, you get $88.20, which would probably be rounded down to $88.

23. If the original price is greater than $100.

24. **a.** 75 **b.** $123/185 \approx 12/18 = 2/3 \approx 67\%$

25. Technically it comes to $305,300, but the more reasonable answer is $300,000, because 115% is an average.

26. $3200/8400 = 0.38$ and $20,400 / 59,000 = 0.35$, so the 1976 purchase took a bigger portion of his income than the 2006 Honda did.

27. An increase of 71%.

28. Previous year $= \dfrac{72.0}{1.092} \approx 65.9$ ppm.

CHAPTER 7 Uncertainty: Data and Chance

SECTION 7.1 Representing and Interpreting Data

Note: In cases where students are asked to explain the graph or to describe questions about reliability, validity, etc., there are many possible valid ways to answer those questions. Because of space limitations, only one response is given here. However, it should be interpreted as "one of many possible valid responses," as opposed to the right response or even the best response.

1. Answers will vary. Possibilities are given here:

   a. How many times can a third grader dribble a ball in one minute – only one hand can be used, the ball must visibly bounce off the floor each time, it can only touch the floor (i.e., not bounce off the wall).

   b. How long can you hop on one foot – person can select one foot but cannot alternate from one foot to the other, everyone must hop with arms at sides or hop with arms out straight, all hop with same footwear the other foot should be in same position, e.g., just off floor or by the knee, etc., can't touch anything with hands, for each hop the foot must be visibly off the ground.

   c. How many concerts have you attended in the past year – can include free or paying, just focused on music, a street festival counts as one concert.

   d. How much time do you study in a week – include all time spent on your courses outside class, reading, studying, working on projects. Round to the nearest hour.

3.

```
   x  x
   x  x
   x  x
 x x x x
 x x x x    x
 x x x x  xx  xx              x       xx    x
 1 1 2 2 2 2 2 2 2 2 2 3 3 3 3 3 3 3 3 3 4 4 4 4 4 4 4
 8 9 0 1 2 3 4 5 6 7 8 9 0 1 2 3 4 5 6 7 8 9 0 1 2 3 4 5 6
```

   a. There were 27 students whose ages range from 18 – 46. The cluster, from 18 – 21, contains 18 students or 2/3 of the class. Only 4 students are older than 27.

   b. 24. That feels like the center of gravity, as if all these x's were on a see-saw.

   c. mean 24, median 21, modes 19 and 21

5. a. Mean is 123, median is 122, and mode is 121.

   b. They do not tell us anything about the shape of the data—the range, the spread, gaps, clusters, or outliers.

   c.

```
                              x
                              x
                              x  x
                           x  x  x              x
  x                        x  x  x  x    x x    x x
  x            x     x      x x x x x x x x x x x x   x x      x      x
 107   109   111   113   115   117   119   121   123   125   127   129   131   133   135   138
```

d. The number of raisins range from 121 to 137; most lie between 121 and 128.

e. It does not tell you the average.

f.

100	2
110	6
120	25
130	5

g.

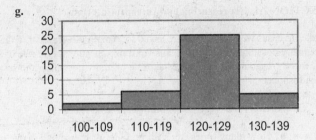

h. The histogram tells us that most of the boxes had between 120 and 129 raisins.

i.

100	0
105	2
110	2
115	4
120	16
125	9
130	2
135	2

j.

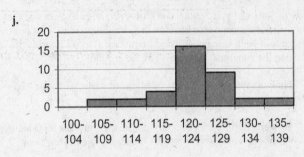

k. The histogram tells us that most of the boxes had between 120 and 129 raisins.

l. They are alike in that they both show a spike in the 120s. They are different in that the second histogram shows that the largest concentration is in the 120-124 range and that, other than the 120s, the number of raisins in each interval is relatively constant.

m. The boxplot tells us that the range was from 107 to 138 raisins, that half had less than 122 raisins, that half had between 120 and 127 raisins.

n. Answers will vary.

7. a. Answers will vary.

b. The mean is 4.2 years; the median is 4 years.

9. 93

11. Let x be the number that is removed. $\dfrac{4(7)+x}{5}=6 \Rightarrow 28+x=30 \Rightarrow x=2$.

13. a. Mean is 72.75 and median is 77.5.

b. Delete one below and one above the median.

c. Delete two below the median.

d. Delete two scores whose sum is greater than 146.

e. Delete one score above and below the median; the sum of the scores must be greater than 146.

15. a.

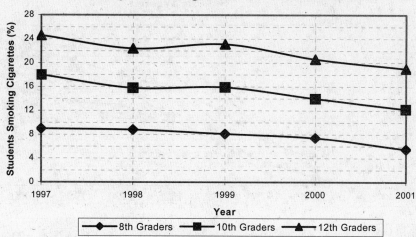

b. For all grade levels, the percentage of students that are smoking cigarettes has decreased over the last 12 years.

c. 8th grade; decrease of ≈ 68%

10th grade ; decrease of ≈ 56%

12th grade ; decrease of ≈ 43%

d. Answers will vary.

17. a.

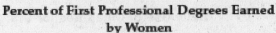

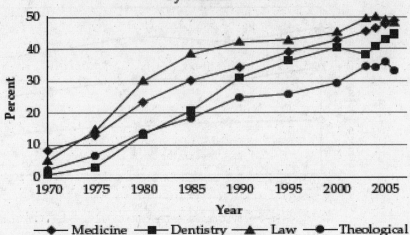

b. Answers will vary.

c. The percentages of women receiving degrees in these fields have risen dramatically in the past 36 years. By 2006 about 4 out of every 10 degrees in medicine and law went to women, and about 3 in 10 theological degrees went to women.

19. Answers will vary.

21. Answers will vary.

23. Answers will vary.

25. a. The mode is New Hampshire. The mode is the datum that occurs the most.

b. There is no mean. Although you can, and computer programs will, compute the numbers, the number 4.6 is meaningless.

27. a. Mean is 44/12 = 3.7, median is 2.

b. There should be spaces between the bars representing 6, 9, and 11 siblings. Without those spaces, one could interpret the data as being clustered together.

29. Answers will vary.

31. There are at least a few employees making substantially more than $10 an hour.

33. Think of a set of data where the mean and median are close and add one more datum which is an outlier on the high end. The median will either stay the same or shift to the next highest datum, while the mean will jump appreciably. If you have outliers, you can have a situation where the mean is not really close to the center of the data. See the example below for wages (in thousands of dollars) in a small company:

12, 12, 12, 15, 15, 15, 18, 18, 18, 18, 18, 18, 20, 20, 20, 30, 30, 30, 75, 100

In this case, the median is 18,000. The mean is 26,000 and ¾ of the employees make less than 26,000. That is, ¾ of the data are below average.

35. Answers will vary.

37. Rates allow us to compare different wholes. For example, say there were 20 murders last year in a city of 800,000 and 30 murders in a city of 1,800,000. If we just show the raw data, and the reader did not know the populations of the city, one could conclude that the second city was more dangerous. However, if we use 100,000 as our unit, we would say that the first city has a murder rate of 2.5 per 100,000 people and the second city has a murder rate of 1.7 per 100,000 people.

39. What is crucial is that there is a whole. Many data can be in percentages but there is no whole. For example, see problem 23.

41. a.

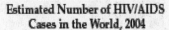

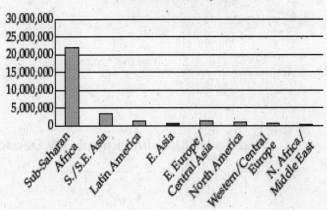

b. The number of HIV/AIDS cases in Sub-Saharan Africa clearly stands out on this graph.

c. Answers will vary.

43. a.   Answers will vary.
  b.

Number of U.S. Immigrants by Decade

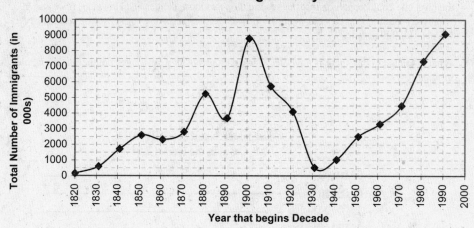

Number of U.S. Immigrants by Decade

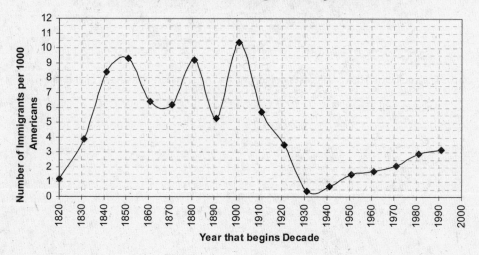

The first graph makes it appear that there has been a rapid increase in the number of immigrants in the last few decades. The second graph gives a more accurate impression that the <u>rate</u> of increase is actually not as high as it has been in the past.

c.   Answers will vary.

d.   Answers will vary.

45. Answers will vary.

47. Answers will vary.

49. Answers will vary.

51. a. If you enclose each column of x's in a line plot with a bar, you have a histogram.

   b. If you put a dot at the center of the top of each bar and connect the dots, you have a line graph.

53. 7

55. 5 more votes

57. D

SECTION 7.2 Distributions: Centers and Spreads

1. **a.** It tells us the range: 52 – 81 and that there is a cluster in the mid to high 60s. We can say that about ½ the class has a pulse rate between 65 and 69.

```
                         x x x
x            xx    xx   xxxxx        xx          x
5 5 5 5 5 5 5 5 6 6 6 6 6 6 6 6 6 6 6 7 7 7 7 7 7 7 7 7 7 8 8
2 3 4 5 6 7 8 9 0 1 2 3 4 5 6 7 8 9 0 1 2 3 4 5 6 7 8 9 0 1
```

    b. It tells us the range and that ost of the pulses are in the 60s.

        5 | 289
        6 | 236778899
        7 | 145
        8 | 1

    c. It tells us the range, that the median is 67, and that about half the class's pulse is between 62 and 70.

      6 33
      *
      7
      * 56777
      8
      * 5559
      9 01
      * 58

    d. I would have everyone sitting down. I would make sure that everyone got their pulse in the same way, e.g., finger on the wrist. I would say "1, 2, 3, start" and then "stop" after 30 seconds. Some people don't have the best concentration and so getting the number for 30 seconds and then doubling it would be fine.

3. **a.**

                                4 | 5
                                5 |
                        32 | 6 | 47
          9764444310 | 7 | 02256
              821 | 8 | 012348
                6 | 9 | 48

    b. The first class has a much larger range (45 to 98) compared to (62 to 96). The first class has about the same number of scores in the 70s as in the 80s, while the second class has a large cluster in the 70s.

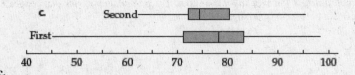

    c.

    d. We can still see that the range for the first class is much greater. We can see that the median for the first class is higher (78 to 74). We can also see that the middle half of the scores are similar (about 71 to 83 in the first class and about 72 to 80 in the second class).

    e. Answers will vary.

5. The second class had a median of 79, as compared with 77 in the first class. The means for both classes were 80. The second class was bimodal at 77 and 85, whereas the mode for the first class was 76. The second class had a smaller range, 65 to 96 compared with 58 to 100 in the first class.

7. a. Grouped frequency bar graph, histogram, or circle graph is appropriate. Also appropriate are box-and-whisker and line plot.

b. Since the distribution is skewed, the mean, median, and mode are not convergent. The median is 21. The data are bimodal at 19 and 21. The mean is 23.9.

c. Range: 18-46
Clusters: 18-21
Biggest gap: between 27 and 37
Outliers: none
Standard deviation: 8.0

9. a.

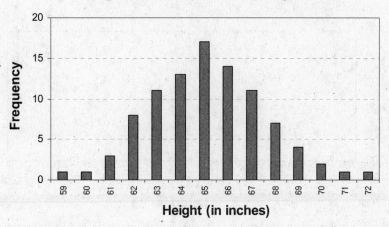

b. Mean: 65.15; standard deviation 2.43.

c. 70%

11. 61, 71. Approximately 2% are less than 5 feet tall.

13. a. 50 tires.

b. Without z-scores, one can approximate the area under the curve. By various means, one can conclude that approximately 6% will wear out before 55,000 miles.

15. a. Answers will vary.

b.

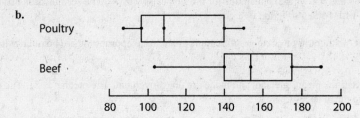

c. Answers will vary.

d. The relationship between calories and sodium content is pretty strong. That is, in general, the more the calories the more the sodium.

d.

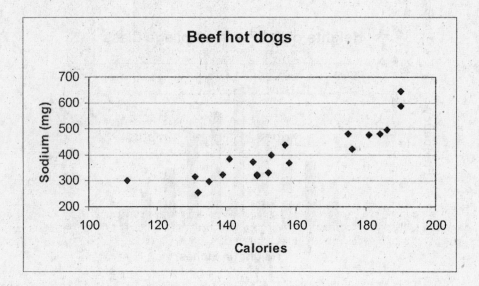

17. a. Positive correlation

b. No correlation

c. Negative correlation

d. Positive correlation

e. Positive correlation

f. Positive correlation

g. Positive correlation

h. Positive correlation

19 – 21. Answers will vary. Below I describe interesting aspects for each set of data.

19. a. Whom did you survey? I would not say that over half of the families I know eat dinner together 5 or more days a week. Does it count if only part of the family is there? Were these data gathered from two-parent families or from one- and two-parent families?

b. How often does your family eat dinner together in an average week during the school year? (I would give them the categories in the table below or I would ask for a specific number. For example, a response of "2 or 3" would create problems in comparing to the data given.)

c.

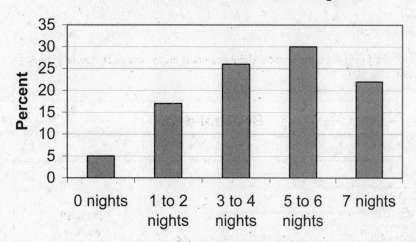

d. A circle graph would have been okay here, but there is a numerical progression (from none to 1 or 2 to 3 or 4 to 5 or 6 to 7), and it is easier to follow this progressiopn with a bar graph. One main advantage of a circle graph is that it gives you the part of the whole; in this case, the data are in percentages, so you get that from the bar graph also.

e. The bar graph does help you see that the percentages increase as the number of days per week eating together increases, up to eating dinner together every day.

21. I would want to know if they gave these categories (0-4, etc.) or asked parents to estimate and then grouped them into these categories. I would want to know if the researchers took a representative sample; for example, kids get more colds in the Northeast than in the Southwest because of the severity of the winters.

23. 76.7

25. Answers will vary.

27. a. For the students: mean is 2.1, median is 1.5 siblings, mode is 1 sibling.
For the mothers: mean is 3.8, median is 3 siblings, the modes are 2 and 3 siblings.

b. In this case, the difference between the two means and the two medians is about 1 ½. So I might say that the students in the class have, on average 1 ½ fewer siblings than their mothers did.

c. We really can't. While we know from the Census that the size of families has been steadily decreasing, the demographics (characteristics) of the students in the class is by no means representative of the overall population.

d. The line plots show that the data for the students is more tightly clustered around 0 to 2 and then steadily decreases after that. The 11 is clearly an outlier.

e. The box plots nicely show the "shift" in data from the mothers to the children. The mothers' median is equal to the students' upper quartile. That is, half of the mothers have 3 or more siblings; only ¼ of the students have 3 or more siblings. The ranges in the data are comparable.

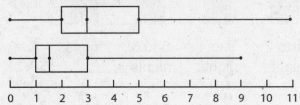

f. The standard deviations are 2.1 and 2.7, respectively.

g. We would need to know the ages of each student and her mother.

h. It would tell us if the size of a student's family is related to the size of the mother's family. That is, do students with small families tend to have mothers from small families and students with larger families tend to have mothers from larger families?

i. Answers will vary.

29. a. Make sure all of the intervals have the same number of years in them.

b. Answers will vary.

c. Answers will vary.

d. The graph from part (a) clearly shows that there are many brand-new teachers in the district—which is not evident from looking at the average.

31. a. The football payrolls range from about 82 to 152 million with an average of about 114 million. The middle half of the teams have payrolls between 100 and 122 million. There appears to be a cluster of payrolls between 100 and 120 million. The baseball payrolls range from about 35 to 210 million with an average of about 82 million. The middle half of the teams have payrolls between 68 and 103 million. The longest whisker is longer than the other whisker and boxes combined which indicates that there are probably gaps between 105 and 210 million and that 210 million might be an outlier.

b. The football distribution is probably slightly skewed to the left. The baseball distribution is strongly skewed to the left.

c. The baseball payrolls have a range (175 million) that is more than the range of the football payrolls (70 million). About ½ of the baseball teams have a payroll smaller than all of the football teams.

33. a. A cluster

b. Answers will vary.

c.

Normal distribution is symmetric.

d.

Skewed to the right means more clustered at the left and more spread out at the right.

35. Without z-scores, one can approximate the area under the curve. By various means, one can conclude that approximately 35%, or about 700 tires, will wear out before 45,000 miles.

37. Her math score was the highest in terms of standard deviations above the mean.

39. Answers will vary.

41. There is a mild positive correlation.

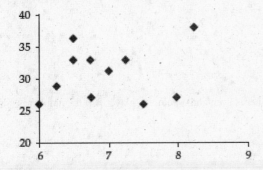

43. a. If we are going to make generalizations about the population called "teenage boys," the driving records of the boys surveyed in these places is likely to be worse than if we had a representative sample.

b. If we are going to make generalizations about the population called "citizens," then parents is not a random sample. Even if we are going to make generalizations about the population called "parents," then parents at a PTA meeting is still not representative. Only a fraction of parents attend PTA meetings.

c. If we are going to make generalizations about the population called "residents," this will not be a representative sample: Not all people have telephones, and certain people are home less than others, for example, young single people.

d. You would probably get different results if you asked their opinion on December 23!

45. Her mean speed is about 17 miles per hour.

47. Julie is right. Although 148.12 grade points / 46 credits = 3.22 GPA, one's total grade points is either a whole number or a mixed number.

SECTION 7.3 Concepts Related to Chance

1. 1/3

3. 0.14 or 1 in 7 ; 0.77

5. **a.** 0.1   **b.** 0.3   **c.** 0.6   **d.** 0.4   **e.** 0.9

7. **a.** ¼   **b.** 3/8

9. **a.** 1/36   **b.** 18/36 = ½   **c.** 15/36 = 5/12   **d.** 6/36 = 1/6

11. 2/3 of a chance of landing in Room A

13. 2/9

15. 1/6

17. No, it is not a fair game.

19. The expected value from the spinner is $1.625. Take the $2 and run. Over the course of 52 weeks, the spinner would yield $84.50, while taking $2 per week would yield $104.

21. 2/3

23. $375.

25. a. It is fair

    b. It is not fair. Give 1 point to player A if the number is even and 3 points to player B if the number is odd

    c. It is fair.

27. **a.** 2/3   **b.** 4/9

29. 1/12

31. 53/80

33. ≈ 0.83

35.

    a.   5/16 = 0.3125

    b.   35/128 ≈ 0.27

    c.   63/256 ≈ 0.25

    d-e. Answers will vary.

37. The most likely sum is 5. The probability of rolling a 5 is 4/16 = 1/4.

39. Answers will vary. $P(3 \text{ doubles in a row}) = \dfrac{1}{6} \times \dfrac{1}{6} \times \dfrac{1}{6} = \dfrac{1}{216} \approx 0.00463$.

41. Answers will vary. Here is one set of possibilities: One die must have the same number on all its faces. The other die could have 1, 1, 3, 3, 5, 5, or three of one odd number and three of another.

43. The probabilities of winning are: player 1: 1/9, player 2: 2/9, and player 3: 6/9
Thus, give each player this number of points when they win: player 1: 6 points, player 2: 3 points, player 3: 1 point

45. No.

47. No. We need to remember the law of large numbers. If you play many times, you are more likely to win about ¼ of the times. But with a small sample size, the unlikely is possible though not probable, like rolling doubles 4 times in a row, for example.

49. Facts: There are at least one red, one blue, and one green ball and there are at least three different colors of balls in the bag. Inferences will vary - there are probably more red than blue (or green) balls; there are probably at least twice as many red as blue (or green) balls; there are probably less than 10 colors, etc.

51. Answers will vary.

SECTION 7.4 Counting and Chance

1. **a.** 1/52

    b. 1/13

    c. 3/13

3. **a.** 4 x 3 x 2, 4!, 24

    b. 16, determined by making the arrangements

5. 10,000

7. 3/51 = 1/17

9. $_{12}C_3 = 220$ ways.

11. If the flavors were scooped in any order, there would be $_9C_3 = 84$ possibilities. If you specified the order of the flavors, there would be $_9P_3 = 504$ possibilities.

13. P(4 of a kind) = 0.0002. P(3 of a kind) = 0.0235. P(2 of a kind) = 0.588.

15. 1/120

17. The probability that each child will get the popsicle he or she wants is $\dfrac{26}{27}$.

19. **a.** 21 choices

    b. 7. (SM, MT, TW, WT, TF, FS, SS)

    c. 6 choices

21. **a.** 39,916,800 possible words from *mathematics*, 11!. Assuming you use all the letters.

    b. Answers will vary.

23. $n!$ can be written as follows: $n(n-1)(n-2)...(n-r+1)(n-r)!$ When $_nP_r$ is expressed as a fraction, $(n-r)!$ in the numerator and the denominator cancel each other, leaving $n(n-1)(n-2)...(n-r+1)$, which is the other expression for $_nP_r$.

25. Answers will vary.

CHAPTER 7 REVIEW EXERCISES

1. a.

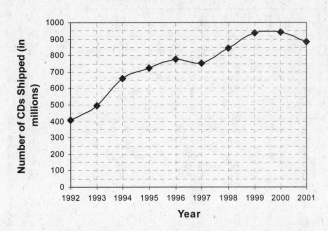

Number of CDs Shipped

b. The number of CDs shipped rose steadily between 1992 and 2001, with a dip in 1997 and 2001. The number of units shipped in 2001 was more than double the number shipped 9 years earlier.

2. a. See graph. Note that because the years are not evenly spaced, we need to leave blank spots for missing years. And we should not connect the dots, because we can't be sure what the data are for the missing years.

b. The percentage of adults who smoke declined pretty steadily between 1965 and 1990. Since 1990, the percentage has declined only slightly, from 25% to 23%.

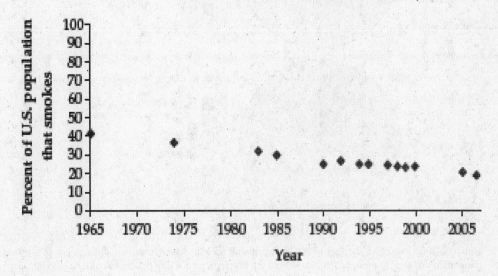

Percent of U.S. Smoking Population

3. a.

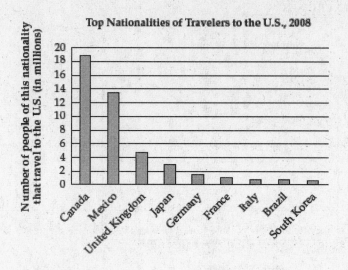

b. Answers will vary.

4. a.

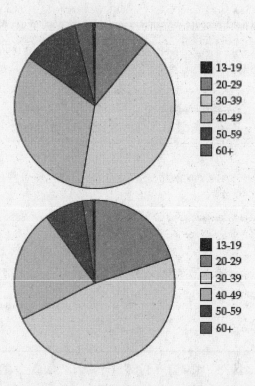

b. Between 1985 and 2000, the ages of AIDS patients increased. An alternative wording: The proportion of AIDS patients in younger age groups decreased, and the proportion of AIDS patients in older age groups increased.

5-7. Responses for these questions will vary.

8. a. One possible graph is below.

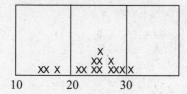

b. The mean is 23.7; the median is 25.

c. The number of drops recorded varied from 15 to 30. Over half of the data were between 21 and 27 drops.

9. a. 4.1 pounds

b. 2.8 pounds

c. Were the dogs all approximately the same size? For example, comparing the weight loss of a 100-pound dog to the weight loss of a 10-pound dog does not make sense.

10. a.

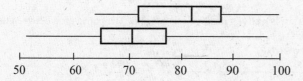

The box plot gives a quick snapshot. It tells us that the first class had a signigficantly higher range; that the second class did better overall - it had a higher median; and that 3/4 of the second class is above 70, compared to only 1/2 of the first class.

b.

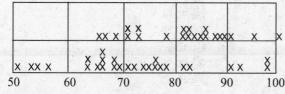

The line plot also lets us see the range; it also lets us see the clusters. The first class is relatively spread out; the second class has a cluster in the 80s.

c. *Box:* pros - quick snapshot; cons - you don't have all the data.
Line: pros - you have all the data; you can see the distribution (spread, range, clusters, gaps); cons - you don't have the average.

d. Means are 72.1 and 80.8; medians are 70.5 and 82.

e. Ranges are 47 vs. 36.

f. Standard deviations are 11.9 vs. 9.3.

g. 71% vs. 70%

11. He needs at least an 83.

12. 48 students.

13. a. 21 **b.** 1.9 **c.** 2

14. a. A situation with outliers.

83

b. A situation where you would want to know standard deviation; mean and standard deviation go together well; grades.

15. a.

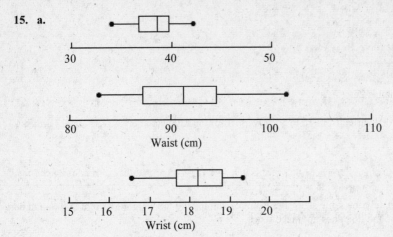

b. The correlations between neck and wrist is a strong positive relationship. The correlation between neck and waist is a weak positive relationship.

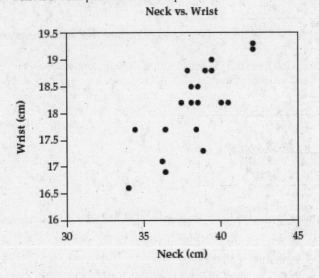

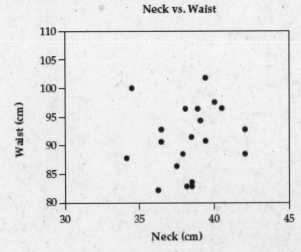

84

16. $\frac{1}{4} \times \frac{2}{8} = \frac{1}{16}$ is the probability of drawing two white circles.

17. a. 1/2

    b. The probability is 1 because there are only two colors.

    c. 5/42 (we can only consider the possibility that there are red socks)

18. Most likely sum is 9, which can be arrived at with 1&8, 2&7, 3&6, 4&5, 5&4, 6&3, 7&2, or 8&1. There are 64 ways to roll the dice, so the probability of rolling a sum of 9 is 8/64 = 1/8.

19. 1/4

20. The probability of randomly choosing a number with a zero is 9/90 = 1/10. The probability of randomly choosing a number with a 5 is 18/90 = 1/5.

21. It's not fair; 3/4 chance of even; 1/4 chance of odd. To make it fair, give 3 points to player A.

22. Not fair. Probability of a match is 1/3. You could make it fair by giving player A 2 points if the colors match and player B 1 point if they don't match.

23. $5! = 120$ ways to display the sports on the poster

24. It doesn't matter what the first card is, there are 12 out of 51 ways to get a matching suit with the second card, thus 12/51 = 4/17.

25. 24

26. 0.2

27. a. 45   **b.** 120

28. a. $_6C_2 = \frac{6!}{4!2!} = 15$ ways to choose a schedule   **b.** 3   **c.** 5

29. 26 x 10 x 10 x 10 x 26 = 676,000 different license plates.

CHAPTER 8 Geometry as Shape

SECTION 8.1 Basic Ideas and Building Blocks

1. **a.** A tetromino is the mathematical name for a Tetris piece.

    b. *Faulty definition:* Four squares that touch each other.
    Fixed definition: Four squares on a sheet of graph paper where each square intersects at least one of the others at a whole edge.

3. **a.** The lines are the same length.

    b. The circles are the same size.

    c. Yes.

    d. The segment on the far right.

    e. Answers will vary. Most likely, a triangle with circles at its vertices will be seen.

    f. It is not a correct 3-dimensional drawing, although it looks like it upon first inspection.

5. 6 different rays.

7. **a.** False. The lines could be skewed.

    b. True. Given any two parallel lines, there is a plane that will contain both lines.

    c. True. The lines contain three distinct sets of points: those exclusively on one line, those exclusively on the other line, and the point of intersection. These points determine the plane in which the lines lie.

    d. True. A line that is perpendicular to a plane forms right angles with all lines in the plane.

    e. False. Think of two lines on the plane determined by this sheet of paper and a third line perpendicular to this plane.

    f. False. Two planes cannot intersect at a point.

9. Several answers are possible. Examples are given.

    a. $\angle ABG$ and $\angle GBE$; $\angle AEB$ and $\angle EAB$. **b.** $\angle FEG$ and $\angle GEB$; $\angle AGB$ and $\angle BGE$.

    c. $\angle ABC$ and $\angle EBD$. **d.** $\angle ABC$ and $\angle ABG$.

11. **a.** 24 times **b.** 9:00 and 3:00 **c.** 150° and (210°)

    d. 75° **e.** 27° **f.** Answers will vary.

13. Answers will vary.

15. **a.**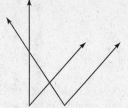
    b.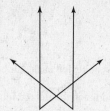
    c.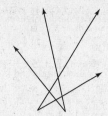

17. Answers may vary. A possible letter "e" for each font is given below.

87

SECTION 8.2 Two-Dimensional Figures

1. Answers will vary.

3. a.   I would say the triangle and kite are most alike, but each pair has similarities.

    b.   There are several properties that are in common to all three figures (these are not underlined). You could make an argument for similarities between the 2nd and 3rd figures or for similarities between the 1st and 3rd figures.

    c.   The 1st and 3rd figure are the most similar.

5. a.   Triangles 20

    b.   Rectangles 7

7. a.
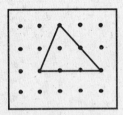 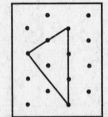

    b.
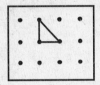

Impossible on Isometric Dot paper; although you can construct a right angle, you cannot construct two equal sides adjacent to a right angle.

    c.
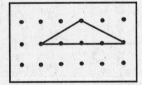

d. Cannot construct an equilateral triangle on Geoboard paper.

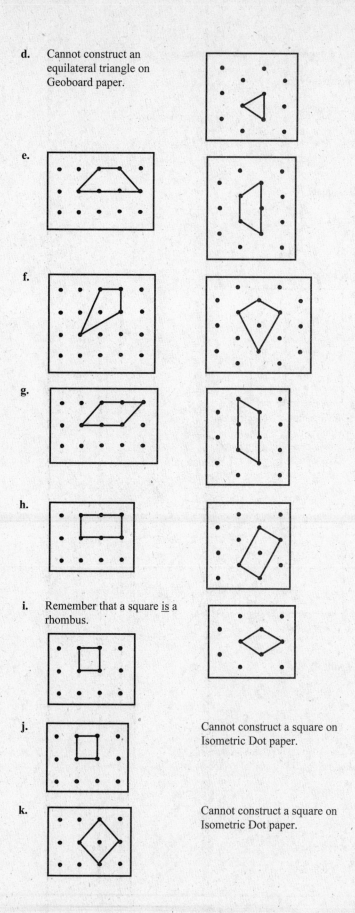

e.

f.

g.

h.

i. Remember that a square is a rhombus.

j. Cannot construct a square on Isometric Dot paper.

k. Cannot construct a square on Isometric Dot paper.

9. **a.** Concave polygon **b.** Not a polygon **c.** Convex polygon

    d. Not a polygon **e.** Concave polygon

11. **a.** Parallelogram, rhombus, rectangle, square.

    b. Rectangle, square.

    c. Square, rhombus.

    d. Parallelogram, rhombus, rectangle, square.

    e. Rectangle, square, isosceles trapezoid.

    f. Rectangle, square.

    g. Parallelogram, rhombus, rectangle, square.

    h. Kite, unless it is a rhombus.

    i. Rhombus.

    j. Rectangle.

13. **a.**

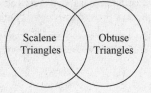

b.

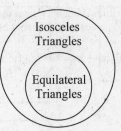

c.

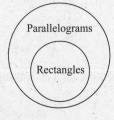

d.

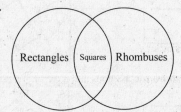

15. A letter "U" (assuming you connect H with A).

17. **a.** $\left(\dfrac{0+4}{2}, \dfrac{4+10}{2}\right) = (2,7)$

    b. $\left(\dfrac{3+7}{2}, \dfrac{4+12}{2}\right) = (5,8)$

19. a. $(0,0)$, $(6,0)$, $(6,6)$, and $(0,6)$

b. The other two vertices are on the horizontal line going through $(7,-2)$ and are equidistant from $(7,-2)$; for example, $(5,-2)$ and $(9,-2)$.

c. There are many possibilities, here are two: The coordinates are $(10,0)$ and $(0,10)$, or $(10,0)$ and $(0,-10)$.

d. Answers will vary.

21. If the definition does not include both parts, then there could be more than one kind of regular *n*-gon for a given length of sides. For example, a regular four-sided figure could be a square or any variety of rhombus. A hexagon could be convex or concave.

23. Answers will vary.

25. a. You can make many different hexagons, both concave and convex, that have all sides equal but not all angles equal. Some have symmetry, some don't.

b. Many possibilities.

c. Many possibilities, all are concave.

d. A few shapes are possible. All have four consecutive right angles. After that, the last two sides can make a shape that looks a bit like Utah, or a shape with an external acute angle, or a shape with a piece "jutting out."

e. Two possibilities: both have five 90° angles and one 270° angle. One kind has reflection symmetry and the other doesn't.

f. No.

g. None.

h. No.

i. Yes, two possibilities.

27. A square is derived from a rhombus and a rectangle, because it has four equal sides that meet at right angles.

29. b. Only one way.

c. Label the angles *a*, *b*, and *c*. At each point where three triangles meet, the angles are *a*, *b*, and *c*. Since we know that the sum of these is 180°, each of those points is a straight line.

b. Large triangle is similar to the small triangle.

31. Note that $n-2$ triangles can be inscribed in any regular *n*-gon originating from the same vertex. All the triangles have angles summing to 180°, so the *n*-gon has angles summing to $180(n-2)°$. For example:

This 6-gon (hexagon) has 4 inscribed triangles and $180 \times 4 = 720°$.

33. **One way:** Uses a straight edge and compass. Draw any line through the circle, call the points where it crosses the circle A and B. Using the compass, find the perpendicular bisector of AB, call the points where this meets the circle C and D (notice that CD is a diameter of the circle). Now bisect CD and the midpoint of CD will be the center of the circle.

35. **a.** A 3RIT is a figure made from 3 right isosceles triangles joined together so that when a side meets a side, there is no overlap.

  b. Here are some examples:

  c. They are not different. Flip vertically to see the congruence.

  d. Here are some examples:

37. $A(0,0)$ $F(8,4)$
  $B(8,0)$ $G(8,8)$
  $C(4,4)$ $H(4,8)$
  $D(6,2)$ $I(2,6)$
  $E(6,6)$ $J(0,8)$

39. **a.** Orient the triangle so that its base is on the x-axis with one vertex at the origin. Let the other two vertices be called $(4a,0)$ and $(2a,2b)$. The midpoints of the isosceles sides are (a,b) and $(3a,b)$. The base measures

$$d = \sqrt{(4a-0)^2 + (0-0)^2} = \sqrt{16a^2} = 4a$$ and the distance of the segment connecting the midpoints is

$$d = \sqrt{(3a-a)^2 + (b-b)^2} = \sqrt{4a^2} = 2a.$$ Therefore the segment is one-half the length of the base.

  b. True for all triangles, can prove the same way as in part (a).

41. **a.** There are many ways to show this. One way is to make use of the Pythagorean Theorem.

  b. One diagonal has slope 1 and the other has slope -1. $(1)(-1) = -1$, so the lines are perpendicular. (Note that this assumes that the square is oriented with its edges running vertically and horizontally.)

43. Answers will vary, but might include: It has four sides, it has more than one right angle, all the sides are congruent, opposite sides are parallel.

45. D

47. a. They don't have the same number of sides and angles

  b. All the sides and angles are congruent.

SECTION 8.3 Three-Dimensional Figures

1. **a.** a cube, a rectangular prism are two possibilities

    b. pentagonal pyramid

3. Answers will vary. There are several correct answers.

    a. $\square ABC$ **b.** Point A **c.** $\overline{AB}$

5. **a.** Octagonal prism   **b.** Triangular prism   **c.** Rectangular prism   **d.** Right cylinder

7. **a.** They have at least one pair of parallel sides.

    b. Prisms have two parallel bases that are congruent polygons.

    c. At least one circular base.

9. Figures on isometric dot paper should look the same as figures in book.

11. **a.** rectangle

    b. rectangle

    c. rectangle

    d. rectangle

    e. rectangle

13. $2n$ vertices; $n+2$ faces; $3n$ edges.

15. Answers will vary.

17. Yes. If all the edges of the base are of different lengths, then the triangular faces will not be congruent.

19.

Base of prism	Number of diagonals
Triangular	$2 \cdot 3$
Square	$3 \cdot 4$
Petagonal	$4 \cdot 5$
N-gon	$(n-1)n$

21. The sides will always be isosceles triangles.

23. There are nine.

25. A

27. Answers will vary.

29. A

31. D

33. C

CHAPTER 8 REVIEW EXERCISES

1. Answers will vary.

2. There are multiple answers for each question. One answer is given for each.

   a. $\angle ABG$ and $\angle GBF$

   b. $\angle CBG$ and $\angle GBF$

   c. $\angle ABG$ and $\angle DBE$

   d. $\angle DBE$ and $\angle EBG$

3. **a.** False. Consider two lines that lie on this paper and one that is perpendicular to the paper.

   b. False. They could be skew.

4. If you do not start with some undefined terms, some of the definitions will be circular, similar to dictionary definitions. For example, A is defined in ters of B, and B is defined in terms of C, and C is defined in terms of A.

5. Answers will vary. One option is to say that a triangle is a shape made by three line segments such that the endpoints of each segment touch the endpoints of another segment.

6. 14

7. Answers will vary. One option is to draw a quadrilateral and one diagonal, thus making two triangles. Knowing that the sum of the angles of a triangle is $180°$ leads to the conclusion that the sum of the angles of the quadrilateral is $360°$.

8. They are convex polygons, with four sides, four right angles, oppostie sides congruent and parallel, and the diagonals congruent and bisecting each other.

9. **a.**   **b.**   **c.**   **d.**   **e.**

10. The case of five right angles is the most challenging. In a hexagon, there are six angles whose sum is $720°$. If five of the angles are right angles, then the sixth angle would be $270°$, which is possible if the hexagon is convex.

11.

The first figure	The second figure
Hexagon	Hexagon
Concave	Convex
Two sides parallel	Three pairs of parallel sides
Two pairs of congruent sides	
4 acute angles	0 acute angles
2 reflex angles	0 reflex angles
0 right angles	2 right angles
0 obtuse angles	4 obtuse angles
3 pairs of congruent angles	3 pairs of congruent angles
1 line of symmetry	2 lines of symmetry
no rotation symmetry	$180°$ rotation symmetry

12. Answers will vary.

13. a. (1) Kiwis are composed of a pentagon with a line segment protruding from one vertex.

(2) The protruding segment is perpendicular to the side the segment would intersect if it were extended.

(3) Kiwis have 2 adjacent right angles, which is equivalent to saying that each Kiwi has 2 parallel sides that meet a third side at 90° angles. The first not-Kiwi has 6 sides. The second not-Kiwi has the protruding segment not perpendicular to the opposite side. The third not-Kiwi has the line segment inside the pentagon; the fourth not-Kiwi does not have 2 adjacent right angles.

b. Figure (a) is not a Kiwi - the protruding segment would not intersect the opposite side if extended. Figure (b) is a Kiwi. Figure (c) is not a Kiwi - the line segment is partially inside and partially outside the pentagon. Figure (d) is not a Kiwi - t does not have 2 adjacent right angles.

14. Answers will vary. One description: Draw a square and then draw the diagonal that begins at the bottom left corner. Extend the base of the square to the right, about one-third the length of the square. Connect this endpoint to the top right vertex of the square.

15. No, because it implies that there are some rectangles that are not parallelograms.

16.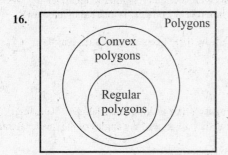

17. ≈ 5.4

18. $(2.5, 7.5)$

19. $(2, 3)$

20. $(6, 6)$ and $(12, 6)$

21. 6 vertices, 9 edges, and 5 faces

22. They have a base and an apex.

23. a. Answers will vary. The top, front, and side views will not work here because two different buildings have the same top, front, and side views. Descriptions can build from the ground up, from one side to another, or from front to back. Also valid is the top view with numbers to indicate the number of cubes in each spot. The key in most cases is to orient the reader correctly.

24.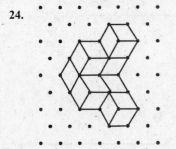

25.

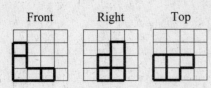

Front   Right   Top

26.

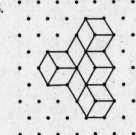
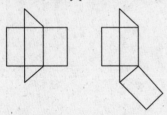

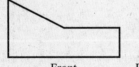

27. There are many possible nets. To be valid, it would have to fold up to make the prism. Two are shown below.

28. Answers will vary. One response: a line segment connecting two nonadjacent vertices.

29. a. Trapezoid

    b. Rectangle

30.

Front   Right-side   Top

CHAPTER 9 Geometry as Transforming Shapes

SECTION 9.1 Congruence Transformations

1. Each point of the figure has been moved 2 units to the right and 3 units down. Also, the figure has been moved approximately 3.6 units along a vector that makes a 55° angle with the *x*-axis.

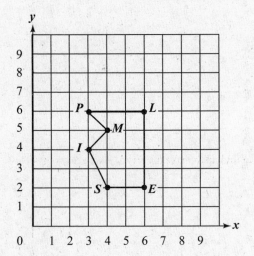

3.

5.

Each point is the same distance from the line as its reflected image.

7.

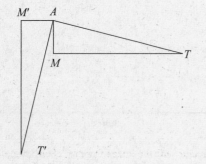

(The original figure is shown for reference purposes.)

9.

11. Place the mirror so that it bisects Molly's right eye and right leg (facing the left side of Molly).

13. There are many ways to do this. One way is to draw a parallel line between $\overline{AT}$ and $\overline{OG}$ and show that the figures are reflections over this line.

98

15. a. Answers will vary.

b. (The original figure is shown for reference purposes.)

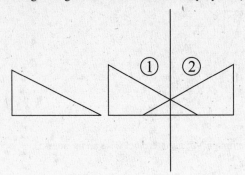

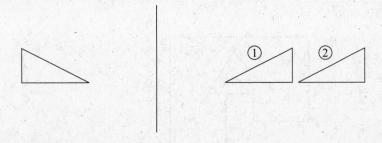

c.

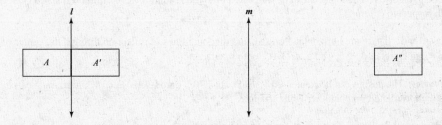

The image from translating and then reflecting does not coincide with the image from reflecting and then translating.

17. Answers will vary.

19. a.

(The original figure is shown for reference purposes.)

b. Answers will vary.

c. The image is located to the left of the original figure by a distance that is approximately equal to the distance between lines *l* and *m*.

99

21. a.

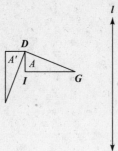 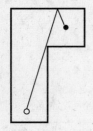 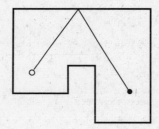

(The original figure is shown for reference purposes.)

b. Reflect the figure across line *l*, then rotate the image 90° counterclockwise about point *D*.

23. a. Reflect $\overline{PR}$ across $\overline{RM}$.

b. Reflect $\square PMR$ across $\overline{RM}$.

25. a.

b.

27. Answers will vary. Some possibilities are given.

a. Baby Blocks: If we see the figure as composed of rhombuses, each rhombus can be mapped onto a neighboring rhombus by a 60° rotation.

b. Broken Windows: The top row can be translated onto the second row. A similar translation can be described with respect to columns.

c. Cross Roads: There are horizontal, vertical, and diagonal lines of reflection through the center of the design, and it can be rotated 90°, 180°, or 270° onto itself.

d. Pinwheel: The design can be rotated 90°, 180°, or 270° onto itself. Each white triangle can be rotated 90° and then translated onto another white triangle.

e. Underground Railroad: The design contains two alternating unit squares that are translated diagonally. One unit square consists of four smaller squares, two light and two dark. The whole design can be reflected across diagonal lines that pass through the center and can be rotated 180° about the center.

29. Answers will vary.

31. a. 9 o'clock.

b. Approximately 6:15.

c. Answers will vary.

33. Answers will vary.

SECTION 9.2 Symmetry and Tessellations

1. **a.** $60°, 120°, 180°, 240°,$ and $300°$ rotation symmetry.

   b.   Five lines of symmetry through the vertices; $72°,$ rotation symmetry.

   c.   Horizontal and vertical line symmetry; point symmetry.

   d.   Translation symmetry if seen as a pattern extending to the right and left.

3. **a.**   Vertical line symmetry.

   b.   Horizontal and vertical line symmetry; $90°,$ rotation symmetry.

   c.   No symmetry because of the line segments.

5. **a.**   Horizontal, vertical, and diagonal line symmetry; $90°$ rotation symmetry.

   b. $90°$ rotation symmetry.

   c.   Horizontal and vertical line symmetry; $180°$ rotation symmetry.

   d.   Horizontal and vertical lines of symmetry.

   e.   Horizontal, vertical, and 2 diagonal lines of symmetry; $90°$ rotation symmetry.

7.

has vertical, horizontal, and two diagonal lines of symmetry.

has vertical and horizontal line symmetry.

have vertical line symmetry.

have one diagonal line of symmetry.

has 90°, 180°, and 270° rotation symmetry.

have point symmetry.

9. Notation after semicolon is from p. 602.

    a. translation; ll

    b. translation, rotation; l2

    c. translation, glide; lg

    d. translation, vertical reflection, horizontal reflection, rotation; mm

    e. translation, rotation; l2

    f. translation, vertical reflection , glide; mg

11. Each arrow shape has three vertex points. Verify that the sum of the angles at each vertex point is indeed 360°.

13. All 12 pentominoes tessellate.

15. **a.** If the unit is one of the C-shapes, then the means of tesselation is a diagonal tranlation to make one strip of C's, then rotate that entire strip 180° and translate. Then repeat these two steps. If the unit is two C's nested into each other, then the means is to translate those two C's diagonally to make a long strip. Then repeat.

b. In this case, four bricks that make a hexagon shape is the only unit.

c. With the interlocking nature of this pattern, there is no unit.

d. In this case, the unit is a hexagon and three consecutive parallelograms that surround the hexagon. It does not matter which threee, but it has to be the same three.

17. It holds.

19. **a-g.** Answers will vary.

h. Impossible

21. **a.**   **b.**

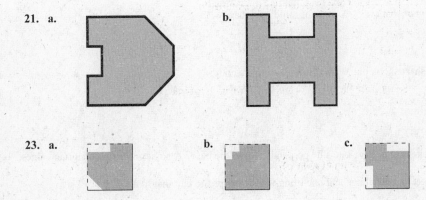

23. **a.**   **b.**   **c.**

25. Answers will vary.

27. All three have the same symmetry: translation, vertical reflection, horizontal reflection, rotation; mm

29. **a.**   90° rotation symmetry

b. The center design has 60 degree rotation symmetry.

c. Depends on what block is picked.

31 - 33.   Answers will vary.

35. Answers will vary. Possible responses: MATH, WAIT.

37. Answers will vary.

39. All the shapes tesselate.

41. **a.**   False. Justifications may vary.

b.   False. Justifications may vary.

SECTION 9.3 Similarity

1. **a.** 13.5 cm

    b. 2.2 cm

3. Yes, the angles are the same and the sides of the figure to the right are twice as long as the sides of the smaller triangle.

5. **a.** 8×11 and 4×6 rectangles are not proportionate.

    b. Answers will vary.

7. 1 inch = 275 miles.

9. Find the height of an object, the length of its shadow, and the length of the shadow of the building. Then write and solve the proportion:

    $$\frac{\text{Height of object}}{\text{Length of shadow}} = \frac{\text{Unknown height of building, } x}{\text{Length of building's shadow}}$$

11. Answers will vary.

13. **a.** The smaller congruent shapes look like the scales on a reptile's body.

    b. Answers will vary.

    c. Answers will vary.

    d. Must be either a parallelogram or have one side perpendicular to the base. Also, it needs two congruent sides.

    e. There are at least two rectangular tilings and one tiling with a composite unit that looks like a "+" sign.

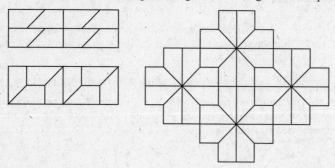

15. It appears that triangles B, C, and E are all similar.

CHAPTER 9 REVIEW EXERCISES

1.

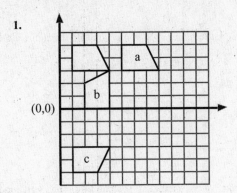

2. Answers will vary.

3.

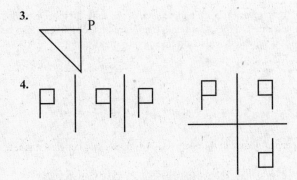

4.

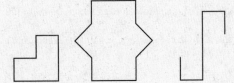

5. The only one-step solution is to rotate the figure 90° counterclockwise, through the point that is 3 units to the right and 3 units below the bottom right vertex of the top trapezoid. There are several two-step solutions; for example, translate 6 units down and then rotate 90° counterclockwise.

6. Alike – they both move the object to a different position, keep the figure the same size and the same shape. Different – the orientation is changed in a refletion but not in a rotation. For example, imagine rotating and reflecting the letter P. To see the rotated P, you could just turn the page until it was "right" again. However, no matter how you turned the page, the reflected P just wouldn't look right – it would feel backwards. The curved part is now facing left instead of right.

7. By including glide reflection, we have the theorem that any figure can be moved to any other spot on a plane in *exactly* one move.

8. **a.** 1/5 turn symmetry, or 72° rotation symmetry; 5 lines of reflection

 b. 1/4 turn symmetry, or 90° roation symmetry; 4 lines of reflection

 c. 1/2 turn symmetry, or 180° rotation symmetry; no reflection symmetry.

9. There are many possibilities. One valid figure is given for each case.

105

10. Answers will vary.

11. a. It means that there is at least one way to pick up the figure, turn it some amount, and lay it down so it fits back exactly on itself.

    b. It means that there is at least one line that you can draw through the figure such that if you fold the figure along that line, the half of the figure on one side of the line will fit exactly on the other half.

12. The symmetries can be described with notation or visually.

    a. ml

    b. ll

    c. mg

    d. lg

13. a. There is more than one possibility. For example, the figure at the left can be seen as the unit. In this case, the means by which it is repeated is a translation to the right. On the other hand, the figure on the right can also be seen as the unit. In this case, the means by which it is repeated is also a translation to the right.

 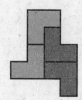

    b. In this case, the unit consists of the five chevrons shown at the right. The means by which it is repeated is translation in a diagonal direction.

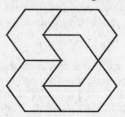

14. a. This figure tessellates. The sum of all angles at every vertex point is $360°$; this is because all angles are $90°$ or $270°$.

    b. This figure tessellates because all quadrilaterals tessellate.

    c. This figure tessellates because it is made by beginning with a square, modifying the bottom, and then translating that modification to the top side.

    d. This figure tessellates. Because of the symmetry of the hexagon, the four acute angles are congruent and the two reflex angles are congruent. If we label the acute angles a and the reflex angles b, we have that $4a + 2b = 720$, which simplifies to $2a + b = 360$. Thus the sum of the angles at each vertex point is $360°$.

    e. This figure tessellates. The reasoning here is virtually identical to the reasoning for part (d), except that the sum of the angles of a pentagon is $540°$.

15. 28 cm

16. Yes. No matter what the size, the ratio between side and hypotenuse is the same.

17.

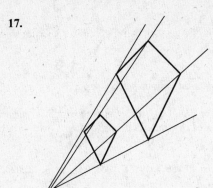

18. $(6,8)$, $(16,8)$, and $(8,14)$. The coordinates of each point of the similar triangle will be double the coordinates of the corresponding points on the original triangle.

CHAPTER 10 Geometry as Measurement

SECTION 10.1 Systems of Measurement

1. Answers may vary. Following are examples for the given objects.

   a. surface area, volume, amount of pollutants, temperature of the water (at various levels), depth

   c. height, surface area of sides (for painting), surface area of roof (for shingles), surface area of windows, surface area of floors, ratio of area of windows to area of floors, ratio of area of windows to area of floors (to determine adequacy of ventilation)

3. a. 0.5   b. 45   c. 0.670   d. 3600   e. 0.450   f. 35,000   g. 0.024

5. *Mental math:* Answers will vary.

   One possible response: Each yard is a little less than $3\frac{1}{2}$ inches less than a meter, so 400 yards would be about $3\frac{1}{2}$ inches $\times 400 = 1400$ inches less than 400 meters. Then notice that 1400 inches ≈ 1440 inches. Since 1440 inches $= 120$ feet $= 40$ yards, we would expect the answer to be a few yards less than 40 yards. *Actual calculation:* ≈ 37 yards.

7. The first estimate is "off" by 5 feet out of 105 feet, which is a little less than 5%. The second estimate is "off" by 5 feet out of 15 feet, which is about a 33% difference.

9. 0.62 and 1.6 are multiplicative inverses. Their product is 1.

11. 30 days

13. Answers will vary.

15. a. The makers of the table essentially set the American/London foot as the unit, 1000 parts, and the other numbers in this column enable us to see the length of other "foots" in relation to the American/London foot. By setting the unit foot as 1000 parts, they were able to avoid decimals.

    b. A bar graph would show the relative lengths of 1 foot in other countries. It could be done alphabetically for easy reference, or in increasing or decreasing order.

    c-e. Answers will vary.

17. a. Answers will vary.

    b. The difference would be about 3/4 gallon.

19. Answers will vary.

21. The answer cannot be determined unless you know the length of the train.

23. Same amount.

25-29. Answers will vary.

31. Answers will vary.

33. 24 glasses

35. A

37. C

SECTION 10.2 Perimeter and Area

1. **a.** $P = 69$ mm, $A = 195$ mm^2

    b. $P = 80$ mm, $A = 148.5$ mm^2

    c. $P = 6$ cm, $A = \sqrt{3} \approx 1.7$ cm^2

    d. $P \approx 37.68$ cm, $A \approx 113.04$ cm^2

    e. $P = 40$ cm, $A = 72$ cm^2

    f. $P = 35.8$ cm, $A \approx 51.9$ cm^2

3. **a.** 15 square units   **b.** 19 square units   **c.** 29 square units

5. Approximately 11.2 feet. Arc length $= \dfrac{128}{360}\pi(10)$.

7. Strategies will vary.

    a. 175 square centimeters

    b. 180 square centimeters

9. 144 sq. in.

11. About 1 foot.

13. Approximately 45.84 square meters.

15. 44.4% = 4/9 of the square.

17. **a.** Answers will vary. The curved line traces the top half of one circle and the bottom half of another circle. These circles have their centers along a diameter of the larger circle and radii that are half the radius of the larger circle.

    b. πr

19. 312 bricks.

21. The diameter of the tree is approximately 16.4 feet.

23. Answers will vary.

25. **a.** Answers will vary.

    b. The area of each piece is 2 square units.

27. Answers will vary according to the person's height. If you assume 1 quarter is 2.5 cm long, the value of a line of quarters 170 cm long is $17. If you assume 1 nickel is 2 mm tall, then the value of a stack of nickels 170 cm high is $42.50.

29. 78 triangles.

31. No, it will not fit.

33. 24,450 miles. 489 miles is $\dfrac{7.2}{360}$ of the earth's circumference.

35. Answers will vary.

37. Answers will vary.

39. This is counter-intuitive: Let d be the diameter of the ball. The height of the can is $3d$, but the distance around the can is $3.14d$!

41. Ratio of corresponding sides is 2:1. Ratio of areas is 4:1.

43. **a.** 10π cm

b. Not possible, because we don't know the height of the part that is cut out.

c. $34 + 3\pi$ inches.

45. **a.** 25 square inches   **b.** Not possible.

c. Answers will vary. If the rectangle is 4 centimeters $\times$ 5 centimeters, $P = 18$ centimeters. If the rectangle is 10 centimeters $\times$ 2 centimeters, $P = 24$ centimeters.

d. 180 feet by 90 feet.

47. The rectangle is getting bigger in its length and width. The area of the new rectangle is $1.5^2 = 2.25$ times the area of the old rectangle.

49. This would be about 732 square miles, which is about 1/2 the size of Rhode Island!

51. The computation yields the almost unbelieveable answer of 2.5 square feet per person. If we make a rectangle that is 5 feet long, it would have to be 6 inches wide. This number helps us to understand why as many as 1/3 of the people died during the voyage. Other accounts of slave ships report that each male had a space about 16 inches wide.

53. Answers will vary depending on the density of the grass.

55. **a.** If we assume that each hamburger is 1/4 pound and a weight of 100 pounds, that person could each 400 hambers a day.

b. If we take a cereal box that ha 12 ounces of cereal and divide 100 pounds by 3/4 of a pound, we get 133 boxes of cereal!

57. 8 cm by 4 cm

59. C

61. B

SECTION 10.3 Surface Area and Volume

1. **a.** S.A. = 1474.3 square inches.
         V = 2160 cubic inches.

    b. S.A. 35,630 square feet.
         V = 312,000 cubic feet.

    c. S.A. 242 square feet.
         V = 267 cubic feet.

3. **a.** S.A. = 144π or 452.16 sq. in. ; V = 288π or 904.32 sq. in.

    b. S.A. = 48π or 150.72 sq. in. ; V = 47π or 150.72 cu. in.

    c. h = $16 / \pi$ or 5.1m

5. **a.** S.A. = 47 square feet.
         V = 16 cubic feet.

    b. S.A. = 695.2 square feet
         V = 1221 cubic feet

    c. S.A. 1980 square feet.
         V = 2985 cubic feet.

7. **a.** S.A. = 216 sq. in. ; V = 144 cu. in.

    b. Make it 6 in by 6 in by 4 in, S.A. = 168 sq in.

    c. Make it a cube. Length of each side would be 5.24 in, S.A. = 164.75 sq. in.

    d. Answers will vary. My guess is that people are conditioned to want taller, skinnier packages.

9. If you roll it so the height is 11 inches, the volume is 62.9 cu. in. If you roll it so the height is 8.5 inches, the volume is 81.7 cu. in.

11. **a.** 60 feet

    b. 44.8 feet

    c. 34.6 feet

13. **a.** Predictions and explanations will vary.

        The new tent is about 291 cubic feet larger than the old tent.

    b. Predictions and explanations will vary.

        The new tent uses almost 258 square feet of additional material.

    c. Answers will vary.

15. 105,000 sheets. The shelves can hold about 42% of the yearly purchase.

17. $\approx$ 34 feet.

19. **a.** Stack them in two layers with each layer a square, 2×2.

    b. Stack them end to end in one layer.

21. Answers will vary.

111

23. **a.** Prediction: $2^3 +$ (4 half cubes/face $\times$ 3 faces) $+ 1$ on the edge $= 8 + 6 + 1 = 15$ cm^3.

   b. 2.54^3 cubic centimeters ≈ 16.387 cubic centimeters.

   c. Answers will vary.

25. **a.** S.A. of small prism : S.A. of large prism :: 52 : 468 :: 1 : 9. The surface area is 3 times larger in 2 dimensions.

   b. Volume of small prism : volume of large prism :: 24 : 648 :: 1 : 27. The volume is 3 times larger in 3 dimensions.

27. 20 centimeters $\times$ 20 centimeters $\times$ 10 centimeters.

29. **a.** Answers will vary.

   b. They would take about the same number of bricks, but the U-shaped one would take more because of the corners.

   c. 8 gallons, 9.1 gallons, 14.6 gallons.

31 - 33. Answers will vary.

35. Answers will vary. Depends on measured dimensions of a ping-pong ball and how you intend to pack the balls.

37. There are other considerations. A cylindrical shape might require less packing material, but there will be wasted space when many of the product is packed into boxes for shipping to stores.

39. 283 cu. in.

CHAPTER 10 REVIEW EXERCISES

1. Answers will vary.

2. **a.** mm   **b.** l   **c.** ml   **d.** g   **e.** kg

3. **a.** 34,000 cm   **b.** 345 g   **c.** 2750 ml

4. Answers will vary. Two situations are given here.

   a. Measuring the length of a figure with a ruler but beginning at 1 instead of 0, a mistake commonly made by children.

   b. Reporting the distance between Los Angeles and San Francisco to the nearest 100 miles.

5. Attributes include surface area, perimeter, temperature, depth, and average depth.

6. Here are two ways: (1) Find the area of the rectangle that encloses the figure and subtract the area of the two triangles: $42 - 12 - 3 = 27$ square units. (2) Use the symmetry of the figure – the bottom half is a trapezoid with area 13.5 square units.

7. Area $= 6 \text{ in}^2$.

8. The area of the border is 146 ft^2.

9. So the dimension is $31.83 \text{ cm} \times 31.83 \text{ cm}$.

10. 1200 sq. in.

11. 200 cm on a side, which is equivalent to a square that is 2 meters on a side

12. The total area of the dartboard is $\pi \cdot 6^2 = 36\pi$. The shaded region is $\pi \cdot 4^2 - \pi \cdot 2^2 = 16\pi - 4\pi = 12\pi$. So the probability of landing in the shaded region is $12\pi / 36\pi = 1/3 = 0.33$ or 33%.

13. **a.** There is empty space between the pennies, because circles do not tessellate.

    b. 25,600 pennies

14. Answers will vary. One response: We can take any parallelogram, slice off a right triangle from one side, and translate that triangle to the opposite side. The parallelogram becomes a rectangle, whose area we know to be $l \times w$. We know first that the rectangle and parallelogram have the same area, that the length of the rectangle is the same as the base of the parallelogram and that the width of the rectangle is the same as the height of the parallelogram. Therefore, $l \times w$ (of the rectangle) = $b \times h$ (of the parallelogram).

15. Answers will vary. Two responses: (1) Because 12 inches = 1 foot, 144 square inches = 1 square foot. (2) One could also draw a square that measures 1 foot by 1 foot; thus the area is 1 square foot. When we convert to inches, the same square is 12 inches × 12 inches = 144 square inches.

16. It quadruples.

17. $37\frac{1}{2}\%$

18. The area of the figure will be about 8.1 cm^2, or 810 mm^2, which means that the actual area of the pond is 810 square meters.

19. Approximately 32.185 cm.

20. 1 ream of paper measures 8.5 inches by 11 inches by 2 inches. Thus all the forms will take up about 433,000 cubic feet. A cube that measures 76 feet on a side would be able to contain all the paper. In more conventional terms, a building 233 feet by 233 feet by 8 feet would be needed to contain all the paper.

21. **a.** 44.7 feet

    b. 34.6 feet

22. **a.** 88.9 cubic yards

    b. 97.8 cubic yards

23. The surface area is approximately 2000 square feet. For the volume, break the figure into a rectangular prism and a triangular prism: 9600 cubic feet.

24. 25,600 squares.

25. There are many examples. Two are $15 \times 1 \times 1$, with a surface area of 62 square units and a volume of 15 cubic units; and $5 \times 3 \times 2$, which also has a surface area of 62 square units but a volume of 30 cubic units.

26. Surface area of the prism is 248 cm^2.

    Surface area of the cylinder is 225 cm^2.

    Volume of the cylinder is 246 cm^3.

    Volume of the prism is 240 cm^3.

27. **a.** 5840 gallons

    b. 1 gallon = 0.1337 cubic foot. The pool holds 5840 gallons, which converts to 781 cubic feet. Since the swimming pool contains 3600 cubic feet, we would fill about 1/5 of the pool.